The Road to Discovery

A Short History of Cold Spring Harbor Laboratory

ALSO FROM COLD SPRING HARBOR LABORATORY PRESS

Illuminating Life: Selected Papers from Cold Spring Harbor, Volume 1 (1903–1969)

Life Illuminated: Selected Papers from Cold Spring Harbor, Volume 2, 1972–1994

The Writing Life of James D. Watson

Grounds for Knowledge: A Guide to Cold Spring Harbor Laboratory's Landscapes & Buildings

The Road to Discovery

A Short History of Cold Spring Harbor Laboratory

JAN A. WITKOWSKI

COLD SPRING HARBOR LABORATORY PRESS
Cold Spring Harbor, New York • www.cshlpress.org

The Road to Discovery
A Short History of Cold Spring Harbor Laboratory

Publisher and Acquisition Editor	John Inglis
Director of Editorial Development	Jan Argentine
Project Manager	Maryliz Dickerson
Permissions Coordinator	Carol Brown
Production Editor	Kathleen Bubbeo
Production Manager	Denise Weiss
Cover Designer	Mike Albano/Denise Weiss

Cover artwork: "The Lab" in Autumn, by William B. Jonas, © 2013. Oil on canvas, 30"× 60", inspired by photographs by Arthur Brings. Reprinted with permission, Private Collection.

Library of Congress Cataloging-in-Publication Data

Witkowski, J. A. (Jan Anthony), 1947-
    The road to discovery : a short history of Cold Spring Harbor Laboratory / Jan A. Witkowski.
        pages cm
    Includes bibliographical references and index.
    ISBN 978-1-62182-108-3 (cloth : alk. paper)
1. Cold Spring Harbor Laboratory--History. 2. Molecular biology--Research. 3. Biology--Research.
4. Molecular genetics--Research. I. Title. II. Title: Short history of Cold Spring Harbor Laboratory.

    QH322.C65W58 2016
    572'.4--dc23

                                    2015010902
10 9 8 7 6 5 4 3 2 1

All World Wide Web addresses are accurate to the best of our knowledge at the time of printing.

For a complete catalog of all Cold Spring Harbor Laboratory Press publications, visit our website at www.cshlpress.org.

Contents

Foreword

Multiple events converged in the 1800s to prompt the establishment of a center for research and education in Cold Spring Harbor. The discovery of terrestrial oil in the United States and the growth of the oil industry during the last half of the 19th century ensured the demise of the whale oil industry, long a mainstay of Cold Spring Harbor enterprises, leaving the western shore of the harbor somewhat of a ghost town, with "industry" buildings and houses that needed repurposing. The establishment of teaching laboratories such as the Marine Biological Laboratory at Woods Hole in Massachusetts in 1888 and the Biological Laboratory at Cold Spring Harbor in 1890 provided the scientifically curious with rich marine environments to study, assisted by improvements in microscope lenses that allowed visualization of organisms and cells at higher resolution than was previously possible. Thanks to the foresight and generosity of the Jones family, particularly John Devine Jones who provided the land and rudimentary structures for investigation until a laboratory could be purpose-built, the Biological Laboratory at Cold Spring Harbor prospered because of its proximity to densely populated areas including New York City. At both Woods Hole and Cold Spring Harbor, developmental biology and the comparative anatomy of organisms were prominent in the first course offerings, leading eventually to a focus of research on embryology that lasted at Woods Hole for more than a century.

But research and education at Cold Spring Harbor took a dramatically different direction. Major insights into quantitative genetics were published in 1900 that were, in fact, rediscoveries of the laws of inheritance worked out by the Austrian monk Gregor Mendel some 34 years before. The director of the Biological Laboratory at the time was Charles Davenport, a Harvard-trained biologist and later eugenicist. His vision of merging Darwinian ideas on natural selection and evolution with Mendel's analysis of quantitative traits created the first institution in the United States to focus solely on the new field of genetics. Within a very short period, experiments on corn would revolutionize genetics and agriculture and set the stage for Cold Spring Harbor Laboratory to play an important role in both research and education for the next century and beyond, a truly exciting time in the history of science.

As the 125th anniversary of Cold Spring Harbor Laboratory approached, I began discussions with John Inglis, Executive Director of the Cold Spring Harbor Laboratory Press, and Jan Witkowski, Executive Director of the Laboratory's Banbury Center, about the publication of a history of the institution. John's portfolio as a publisher included monographs on the history of science and biographies of major contributors. It was immediately obvious that Jan should write the history. In addition to spearheading the influential Banbury Conferences, Jan has written articles and books on the history of science, particularly in genetics, a field in which he worked prior to his appointment at Banbury in 1987. Moreover, because of the major efforts of Mila Pollock, Executive Director of our Library and Archives, the Laboratory has a collection of original photographs and materials that would prove to be a valuable resource. The result is a compelling book that tells a remarkable story of how an institution evolved and the substantial roles people here have played in the development of the biological sciences for well more than a century.

What was once a whaling factory and part of an adjacent "Gold Coast" estate now houses, together with five nearby sites, a large and vibrant biological and biomedical research and education institution. Events that helped shape the future of the biological sciences happened within a collection of architecturally eclectic buildings on the shores of a long, deep harbor that at one time hosted casinos, hotels, commerce, and shipbuilding. In the 20th century, Cold Spring Harbor Laboratory contributed major discoveries in genetics, made possible the purification and characterization of important hormones, and nurtured the field of molecular biology, all of which changed society in many ways. Because Cold Spring Harbor has long been a place for the gathering for scientists at courses and meetings, discussions at this institution helped shape new initiatives such as the Human Genome Project and the Innocence Project, an amazing program that has saved the lives of those sentenced to death for crimes they did not commit. Many Nobel Laureates have worked at Cold Spring Harbor or have benefited from the now vast scientific course offerings that are taught here. Perhaps uniquely, many young scientists, including myself, have early in their careers presented their data to peers and luminaries in their field at Cold Spring Harbor conferences, including the long-running Cold Spring Harbor Symposium on Quantitative Biology. More recently, programs at the DNA Learning Center for middle and high school students, taught both locally and internationally, have influenced new generations of scientists and, just as importantly, informed the general public of the dramatic advances in genetics and molecular medicine that are impacting everyone's lives.

The story of Cold Spring Harbor Laboratory is also one of dynamic change, of leaders whose vision gave the institution influence far beyond these shores. The recruitment by Milislav Demerec of important scientists such as Barbara McClintock, Evelyn Witkin, Max Delbrück, and Salvador Luria to Cold Spring Harbor in the 1940s changed the field of genetics and influenced generations of later investigators. Jim Watson's rebuilding of the laboratory infrastructure, expansion of its programs, and initiation of a new research direction set the stage for the many roles this institution now plays in U.S. and world science.

The modern Cold Spring Harbor Laboratory has a recognizable culture of collaboration, vitality, and innovation that keeps evolving. As we look toward the next period of research and education, it is imperative to keep lessons from the past ingrained in our minds. The large number of visitors to Cold Spring Harbor each year helps create an intellectual environment that is hard to reproduce elsewhere. The lack of tenure permits the institution a healthy turnover of faculty that allows retention of the very best while enabling recruitment of younger scientists with new ideas that will change the direction of science. The rather flat organization and the presence of talented and dedicated administrative and facilities staff allow scientists to pursue their interests unencumbered by administrative burdens that plague investigators elsewhere. These characteristics of the current Cold Spring Harbor Laboratory reflect the influence of its past leaders and their deep understanding of its culture. This book captures much of what makes Cold Spring Harbor unique and should be a blueprint for the next 125 years.

Molecular biology, now such a powerful and influential endeavor, is helping address many of society's challenges, including the management of health, agriculture, and the environment. But there is still much to do and much to discover. The concentration of Cold Spring Harbor Laboratory's science on cancer, neuroscience, and plant biology and its increasing collaboration with clinical centers positions the institution to continue to tackle important human problems. I anticipate that Cold Spring Harbor Laboratory will develop technologies and ideas for providing food to an ever-increasing world population and addressing neurological and other disorders of an aging population. Genetics will continue to be part of the institution's DNA but a resurgence of research on metabolism, nutrition, and physiology will be needed to help us understand how organisms use their inherited genes to maximize their quality of life. The events and discoveries of the past will continue to shape our future.

Bruce Stillman
President
July 2015

Preface

On July 7, 1890, the Biological Laboratory at Cold Spring Harbor welcomed its first students for an eight-week course in biology. Now, in 2015, we celebrate notable events of the past 125 years—the establishment of the Biological Laboratory and of the Carnegie Institution of Washington's Station for Experimental Evolution (1904), and the growth of Cold Spring Harbor Laboratory (1963). The Laboratory has become one of the world's leading research and educational institutions, and my hope is that the stories of the men, women, and science told in this book give some insight into how this remarkable and in many ways unlikely transformation took place.

This book owes much to two unpublished manuscripts. In the late 1980s Bentley Glass, a scientist who had been chairman of the Cold Spring Harbor Laboratory Board of Trustees from 1967 to 1973 and then an honorary trustee, began to prepare an account of the Laboratory to mark its Centennial in 1990. About a decade later, Nathaniel Comfort, then the Laboratory's science writer and currently Professor in the Institute of the History of Medicine, Johns Hopkins University, undertook a major revision of the Glass manuscript and added new and more recent material. I have made liberal use of the valuable information in the Glass/Comfort manuscript in preparing this book and I am indebted to both authors.

I have also benefited in countless ways from working over the past 25 years with the two most recent Presidents of Cold Spring Harbor Laboratory. Jim Watson made his first visit to Cold Spring Harbor as a graduate student in 1948 and immediately became aware of the special nature and historic importance of the site. His desire to help secure and build up its scientific heritage brought him to the Laboratory as Director in 1968 and his wish to honor the past while planning for the future informed many of the decisions he made as a scientific leader, visionary of campus development, and instigator of numerous educational programs. Jim was succeeded as Director in 1994 by Bruce Stillman, whose sense of the history of the Laboratory is equally acute and whose recall of the details of the recent past, after 35 years of service at the institution, is remarkable. I have learned an extraordinary amount from both of these individuals who have dedicated so much of their lives to making the Laboratory an exceptional place.

A special thanks is also due to John Cairns, first Director of Cold Spring Harbor Laboratory. John's insights and lucid prose, whether in letters to the trustees or in his annual reports and other writings, capture so well the nature of Cold Spring Harbor.

The story of how a small harbor on Long Island's Gold Coast became home to one of the world's great biomedical research institutes is 125 years long and complex, and this small book cannot be a comprehensive account of the origins and development of the Laboratory. Several minutely footnoted volumes would be required to do justice to all that has gone on here over the years—useful for historians of 20th century science perhaps, but they are not the primary audience for this book.

It is instead intended as a celebration of the fact that despite numerous crises over the years, there is now a world-renowned Laboratory to celebrate. The book is for anyone who knows Cold Spring Harbor by having worked or visited here, or who has driven along Route 25A and wondered about the buildings on the west shore of the harbor. I have tried to provide sufficient information to cover the broad story without getting bogged down in detail or, in describing research, getting lost in technical jargon. I may be thought obsessive by those who do not know Cold Spring Harbor Laboratory and superficial by those who do, especially since in a short history there are inevitably omissions. These are more numerous closer to the present, as the range and complexity of the Laboratory's science increased and the number of researchers rose. There is simply not room to mention all the possible research topics and people involved.

The events described are in chronological order, except on occasions when it seemed more sensible to complete a story from its origin to the present time. The account of an institution is that of its people and its environment and I hope the photographs of people and buildings provide context for the history. I have also included images of some of the key documents in the history of CSHL.

At the end of almost every chapter will be found one to three short essays focusing on a scientist and an experiment or result. Many of these discoveries have been superseded by subsequent research but they were important at the time, both for the work of the institutes at Cold Spring Harbor and the development of biological science more generally. As this is not intended to be an academic history, I have not provided footnotes or references. Instead, I have listed books and papers that I hope will provide a starting point for further readings on Cold Spring Harbor Laboratory.

Many people contributed to this project. At CSHL Press, Maryliz Dickerson (Project Manager) kept track of the project, text, revisions, and figures and had an eagle eye for errors while Kathleen Bubbeo (Production Editor) and Denise Weiss (Production Manager) assembled text and figures into the elegant pages you have in your hands. I am also grateful to Carol Brown, Susan Lauter, Joanne McFadden, and Inez Sialiano for their contributions. Clare Clark and Stephanie Satalino in the CSHL Library and Archives were indefatigable in finding items for me in the archive collections. Judy Cuddihy read the manuscript for errors and wrote several of the vignettes, as did Peter Tarr and Jaclyn Jansen—I am very grateful to them having taken on that job.

(Their contributions are indicated by initials at the end of each vignette.) At Cold Spring Harbor Laboratory, Alex Gann, John Inglis, David Micklos, Mila Pollock, David Stewart, and Bruce Stillman read sections of the text. Bruce was particularly helpful with dealing with the science of the last chapter and he graciously wrote the Foreword. That the Banbury Center kept running efficiently through the long gestation of this book is due to the diligence of Janice Tozzo and Pat Iannotti in my office. At the CSHL Press, John Inglis (Executive Director) and Jan Argentine (Director, Editorial Services) made sure that the project moved smoothly.

The striking painting gracing the cover is by William B. Jones, based on many photographs taken by Art Brings, Chief Facilities Officer at CSHL. It was commissioned by Dill Ayres, Chief Operating Officer of CSHL, and I am very grateful to Dill for letting us use it.

Many people helped find illustrations: Jim Childress (Centerbrook Architects), Jim Duffy (Cold Spring Harbor Laboratory), Richard Gelinas (Institute for Systems Biology), Karen Martin and Robert Hughes (Huntington Historical Society), Scott Schultz (Cold Spring Harbor Library), and Ann Skalka. Thank you all.

Finally, many thanks to my wife, Fiona, who sustained me through the process of writing this book even though my preoccupation with it often taxed her good nature.

Jan A. Witkowski
August 2015

Origins: A Seaside Biology School
for Teachers

As biology neared the 20th century, physiology and morphology became its core topics, gradually displacing the descriptive disciplines of zoology and botany. The fashionable questions concerned evolution, addressed through experimentation and quantitative analysis of studies of embryological development and the relation of morphological form to physiological function.

The favored organisms for embryological studies were marine invertebrates, with their small, transparent embryos and rapid development, and as a consequence a number of marine institutes sprang up to provide ready access to this material. Foremost among them was the Stazione Zoologica in Naples, founded by Anton Dohrn in 1872. Elite American

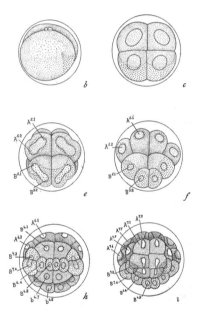

Cleavage stages of the tunicate **Styela partita.**

Stazione Zoologica in Naples.

1

universities, such as Johns Hopkins and Harvard, rented "tables" at Naples, designated laboratory areas for members of those institutions to gain firsthand experience of European-style research.

American institutes followed suit. In 1873 Louis Agassiz, the great Swiss naturalist and director of Harvard University's Museum of Comparative Zoology, founded the Anderson School on Penikese Island in Buzzards Bay off Cape Cod, and Spencer Baird, U.S. Commissioner of Fish and Fisheries, was instrumental in the creation in 1888 of the Marine Biological Laboratory at Woods Hole. Its first director, C.O. Whitman, established a summer program of teaching and research that grew in prominence and soon made Woods Hole the destination of choice for summering American biologists.

A view of Louis Agassiz's biological station on Penikese Island.

The Biological Laboratory Is Founded at Cold Spring Harbor

This period was also a time of great intellectual optimism, before the schism that would later emerge between the arts and sciences. Many institutes, large and small, were founded to foster public interest in advances in science and the arts. The Brooklyn Institute of Arts and Sciences was typical. Founded in 1823, the Institute administered programs and

The Brooklyn Institute of Arts and Sciences.

sponsored work in disciplines as diverse as painting, botany, electrical science, architecture, philology, microscopy, political science, and zoology. Franklin W. Hooper, director of the Brooklyn Institute, had been in the inaugural class at Agassiz' Anderson School for marine biology and had returned to Brooklyn full of enthusiasm for seaside biological field stations. He and Eugene Blackford, a life member of the Institute, began to discuss the possibility of establishing a field station on Long Island. Blackford knew just the spot and in February of 1890, he took Hooper and other members of the Institute out to Cold Spring Harbor further east on Long Island.

Cold Spring Harbor, a harbor flanked by steep bluffs, is one of many bays and coves that form Long Island's intricate northern shoreline. A stream, Cold Spring Creek, enters from the south and a sandbar reaches from the western nearly to the eastern shore, creating a sheltered, shallow inner harbor that empties of water at low tide. In the 1890s, many of the buildings of the village of Cold Spring Harbor, on both shores, dated back to earlier, more prosperous eras of wool and paper mills and whaling.

Eugene Blackford was a New Jersey–born businessman who by the age of 24 had established himself as a fish merchant in Manhattan's Fulton Market. Blackford eventually made a name for himself as an ichthyologist and, despite his lack of formal training, published scientific papers on fish identification and on the decrease in Long Island oysters. In 1879, at the age of 40, Blackford was named the first New York State Fish Commissioner and convinced the Legislature that rearing young trout and other fish in hatcheries and releasing them in Long Island streams and lakes would provide good sport fishing and perhaps commercial harvests that would profit Downstate New York. To finance the project, Blackford approached John D. Jones, a member of a long-established local family whose holdings had once extended as far as the south shore of the Island. Jones agreed to lease the land to the new fish hatchery at a nominal fee and even to provide the use of some old buildings

Franklin W. Hooper, director of the Brooklyn Institute.

Eugene G. Blackford.

The red snapper, Lutjanus blackfordii, *named for Eugene Blackford.*

John D. Jones.

left over from textile manufacturing. The Fish Hatchery at Cold Spring Harbor was established in 1881, in a mothballed woolen mill at the head of the harbor.

Hooper and his delegation from the Brooklyn Institute must have agreed with Blackford that Cold Spring Harbor was a suitable site for their field station because Blackford then contacted his benefactor John D. Jones about the idea. Jones was amenable. He would provide the site; the Brooklyn Institute would appoint a suitable director; and the fish hatchery would cooperate in providing resources for the students. In May 1891, John D. Jones, Walter R.T. Jones, Townsend Jones, Jr., and John J.H. Stewart set up a nonprofit corporation to hold title to his gift of the land, and "to perpetuate the management and care of the grounds and property devoted by him to scientific research." The corporation was called the Wawepex Society, from the Matinecock Indian name for Cold Spring Harbor. The property, consisting of about three acres, was leased for $1 a year to the Brooklyn Institute of Arts and Sciences.

The Biological Laboratory, which opened in 1890, was primarily a summer school for college students, primary and secondary school teachers, and local residents. Courses lasted eight weeks and consisted of hands-on instruction in the natural history of the surrounding harbor, freshwater streams, and other local habitats. Most of the students were women and few had college degrees. The courses gave students a primer in the major types of organisms, exposure to some representative local forms, and training in the basic techniques of natural history and embryology: collecting, microscopy, dissection, and staining.

The Early Programs

The Laboratory's first director was Bashford Dean. Although only 23 years old, Dean was an experienced scientist and in 1890 had just completed his Ph.D. on Paleozoic fish. The following year he was succeeded by Herbert

The first class at the Biological Laboratory.

W. Conn, another instructor, who had directed summer biology laboratories for Johns Hopkins and the Martha's Vineyard Summer Institute.

Conn developed and broadened the Laboratory's teaching program. In his first summer he doubled the Laboratory's offerings by adding a course on bacteriology, and in 1893 he split the course on general biology into systematic zoology and botany. Some residential space was provided for researchers.

The laboratory in the Fish Hatchery building was furnished with laboratory tables, aquaria, hatching troughs, and glassware—as the brochure put it "all the apparatus and appliances required for general biological work," including microtomes, microscopes, and photography equipment. A couple of small boats for collecting marine specimens were moored just yards from the laboratory door.

Herbert W. Conn.

Courses began after the Fourth of July weekend and lasted six to eight weeks, although researchers could come early and stay on until September or later. The 15 students paid $24 each for the course; room and board were an additional $7.50 to $9 per week. Visiting researchers could rent laboratory space for $50 for the summer. Conn also established a series of public lectures for the students and members of the local community. By the Laboratory's third season, these lectures were offered twice a week and attracted 50 to 120 people.

The Laboratory grew rapidly. By 1893, the courses had expanded so much that the Laboratory had outgrown the Fish Hatchery facility. Jones contributed both $5,000 and the site for a new laboratory building across the road from the Fish Hatchery and beside the inner harbor and near Cold Spring Creek.

The new building was ready for use in 1893 and named for Jones himself. The building was 36 feet by 70 feet, large enough to accommodate up to 50 students. A few steps from the door a seawall and small dock were built, affording access to the harbor for collecting trips. An old whaling warehouse was made into a lecture hall and darkroom. In 1895, the Wawepex Society put at the Laboratory's disposal several more abandoned whaling-era buildings. One was converted into a dormitory, later named in honor of Franklin W. Hooper. Adjacent to it was a small cottage also used for a dormitory; in 1898 it became the Director's House, and still later it was named for the biophysicist Winthrop J.V. Osterhout.

The new buildings allowed an expansion of the Laboratory's course offerings. By 1898 it was hosting five courses, all held simultaneously and lasting six weeks. There were four instructors, two assistants, and a total enrollment of 36 students (though a few students took more than one course). A series of evening lectures was offered to the students and interested neighbors.

Funding for the enlarged Biological Laboratory came from two sources: fees paid by the students and "subscriptions" solicited by the

Exterior of the Jones Laboratory, 1895.

THE BIOLOGICAL LABORATORY

— OF —

THE BROOKLYN INSTITUTE OF ARTS AND SCIENCES,

LOCATED AT COLD SPRING HARBOR, LONG ISLAND.

FOURTH SEASON.

ANNOUNCEMENT FOR THE SUMMER OF 1893.

THE LABORATORY BUILDINGS AND EQUIPMENT.

The Facilities for Biological Work at the **Summer Biological Laboratory** of the Institute will be materially increased this season by the erection of a new and commodious laboratory building (36x70 feet) designed for the special purposes of the school. This laboratory building will be built upon a wharf close to the water and will be provided with all the necessary conveniences for summer work. It will contain: (1) a general laboratory with many windows and with tables for students' work, in which there will also be aquaria supplied with running fresh and salt water, blackboards and other conveniences for lecturing and class instruction; (2) a number of small private laboratories, which will be assigned to persons desiring them and who are competent to carry on independent work, and who will, as a rule, be engaged in special investigation; (3) a room equipped for and devoted to work in bacteriological technique, such as making cultures, isolating species of bacteria, etc.; (4) one room equipped with apparatus for photographing purposes, including ordinary photography, microscopic photography and the making of lantern slides; and (5) a working library placed at the disposal of the members of the School. In addition the Laboratory will be furnished with all the necessary apparatus, reagents, etc., needed for biological work at the sea-shore. The Laboratory owns a launch provided with collecting apparatus of all the different forms required for the proper collection of material for laboratory work, and smaller rowboats are also at the disposal of the school. Near by the main laboratory is a second building equipped and used for lecture purposes in cases where larger numbers attend than the general laboratory room will accommodate, or where it is desired to use the lantern for illustrative purposes. The New York State Fish Commission has again shown its generous hospitality to the Laboratory by placing at the disposal of the school portions of the Fish Commission building.

THE PURPOSES OF THE SCHOOL.

The Objects of the Laboratory of the Brooklyn Institute are (1) to furnish a place for general biological instruction and (2) to offer opportunity for investigation to advanced students. The first object to which the energies of the school are devoted is to develop a first-class school of biological instruction to summer students who feel the need of practical study at the sea-shore and of assistance in their work. For this reason the school at Cold Spring Harbor is especially

1893 announcement of the fourth session at the Biological Laboratory of the Brooklyn Institute of Arts and Sciences at Cold Spring Harbor.

Interior of the Jones Laboratory, 1895.

Director from local neighbors to support summer researchers. In 1895, seven generous neighbors contributed between $10 and $100, for a total of $245, presaging events 20 years later when the local community assumed responsibility for the Biological Laboratory (see Chapter 5).

With the use of these old warehouses and dorms, still ripe with the smells of old whale oil, rum, and salt, the western shore of Cold Spring Harbor became once again a little village—not of milling or whaling, but of biology.

Charles B. Davenport Comes to the Biological Laboratory

In 1897, Herbert Conn's impending resignation prompted the Laboratory's Board of Governors to seek a new Director to strengthen its program. Conn recommended Charles B. Davenport, an instructor in zoology at Harvard University.

Davenport was 32 years old. He had grown up just across Long Island Sound, on Davenport Ridge, near Stamford, Connecticut. Davenport earned his A.B. in 1889 and completed his Ph.D. in 1892. He had been

STUDYING FORMS OF LIFE

WORK OF THE COLD SPRING HARBOR LABORATORY.

THE RECENT SESSION OF A SUCCESSFUL BRANCH OF THE BROOKLYN INSTITUTE OF ARTS AND SCIENCES— STEADY EXTENSION OF THE FIELD OF INSTRUCTION.

The third session of the Biological Laboratory of the Brooklyn Institute of Arts and Sciences at Cold Spring Harbor, L. I., which has just closed, was the most successful in its history. The laboratory was organized in 1890. It owed its inception to the Brooklyn Institute, recently organized on a broader basis, and its foundation as a branch of that institution of popular instruction was due to the generosity of John D. Jones and the New-York Fish Commission.

Mr. Jones at the outset contributed a considerable sum of money toward purchasing the equipment of the laboratory, and the New-York Fish Commission offered to the school the use of its buildings and appliances at Cold Spring Harbor.

A conspicuous friend of the school was Eugene G. Blackford, for many years a Fish Commissioner of the State and prominent in the work of pisciculture. Others interested were Prof. Franklin W. Hooper of the Brooklyn

Article about work at the Cold Spring Harbor Laboratory from The New York Times, *September 9, 1892.*

appointed an instructor at Harvard in 1891 and had made a reputation for "quantitative biology," a style of biology intended to make a break with 19th-century natural history and to form a bond with physics and chemistry. His course and book *Experimental Morphology* greatly stimulated such future leaders in biology as William Ernest Castle, Herbert Spencer Jennings, Herbert Vincent Neal, and Walter B. Cannon.

Charles B. Davenport in 1898.

Davenport's wife, Gertrude Crotty Davenport, was also a scientist. She had a bachelor's degree in biology and had been an instructor at Kansas University. She and her husband carried out research together, coauthoring several papers. After moving to Cold Spring Harbor, she not only taught the important course in microscopical methods but quickly became the real manager of the Biological Laboratory itself. Davenport's boundless enthusiasm often led to overimaginative projects, but Mrs. Davenport was an excellent tempering influence. Davenport not only relied on her efficiency and good advice, he loved her deeply and, whenever away from home, wrote to her affectionately every day.

Davenport was familiar with the summer program at the Biological Laboratory, having taught two courses there in 1891. He accepted the summer position, and in 1898 began a 36-year tenure at Cold Spring Harbor.

CHARLES DAVENPORT PROMOTES QUANTITATIVE BIOLOGY

Summary

Charles Davenport was one of the leading proponents of quantitative and statistical approaches to biology, in contrast to the merely descriptive. The two papers discussed here, both reporting research carried out at Cold Spring Harbor, exemplify the types of analytical research that Davenport advocated.

The Research

In 1898, when Davenport was appointed Director of the Summer School at the Biological Laboratory, he was a biometrician and, in his vice-presidential address to the Zoology section of the American Association for the Advancement of Science meeting on Denver in 1901, he "… was convinced that statistical studies are first of all necessary to lay the foundations of the science [of the study of variation]." He used two examples of his own research, conducted at Cold Spring Harbor, to illustrate what he meant.

On coming to Cold Spring Harbor, Davenport was determined to use the available local resources for both teaching and for research on the origin of species. His intention was to determine the geographical distributions of different forms of organisms and examine their adaptations to the differing environments, and he did so on the Cold Spring Harbor sand spit. The sand spit includes the outer beach exposed to the force of the Long Island Sound; the tip, where the current in a channel between inner and outer harbors was strong; and the inner, mud-coated beach. Davenport discussed the relative importance of a species "selecting" an environment to which it has favorable adaptations and the role of selection in driving adaptive changes to a particular environment.

One example he gave was of the mussels found in large numbers at the tip of the sand spit, where their byssus threads, binding them tightly to rocks and stones, were essential for

Charles B. Davenport (ca. 1896), Harvard Zoology.

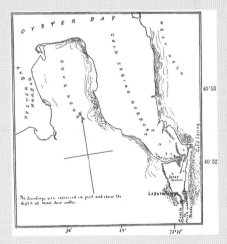

Map of Cold Spring Harbor showing the sand spit separating the inner harbor from the outer harbor (lower right).

Continued

A view of the inner harbor with the south shore of the sand spit on the left.

withstanding the strong currents. He argued that mussels must have had some form of thread—useful, perhaps, for some other function—that enabled them to colonize such environments. Subsequently, natural selection came into play, leading to the survival of variants with longer or stronger threads. Thus, adaptation resulted from a combination of "... the improvement of the organ to meet the requirements of the environment, through selection, and of improvement of situation to meet the abilities of the organism."

Significance

The research in this paper is interesting for what it shows us of the scientific milieu at the time that that Davenport came to the Biological Laboratory. It was not until the 1920s that genetics became the predominant research theme at Cold Spring Harbor.

The Scientist

Charles Davenport began his professional career in the 1890s at Harvard, as an instructor in Zoology. Davenport introduced a new course at Harvard, entitled "Experimental Morphology," that included statistical and experimental studies of variation and heredity as well as physiology. He pursued this approach at length in his book, *Statistical Methods with Special Reference to Biological Variation*, published that year.

At this time, in pre-Mendelian days, studies of heredity were carried out by quantitative measurements of variation and by statistical analysis of the data. The pioneers of this approach were William Bateson, Karl Pearson, and William Weldon—followers of Francis Galton. Pearson and Weldon founded the journal *Biometrika*, and Davenport was its U.S. editor. However, following the discovery of Mendel's work.

Davenport became firmly committed to Mendelian genetics, although he never carried out experimental genetics on, for example, *Drosophila*. He and his wife Gertrude described family studies on the inheritance of eye and hair in humans, but all too soon Davenport turned his attention and energies to eugenics.

— *J.A.W.*

$$2$$

Charles Davenport and the Carnegie Institution of Washington's Station for Experimental Evolution

Davenport and the Biological Laboratory

In 1899, within a year of his engagement as director of the summer program at Cold Spring Harbor, Davenport left Harvard University to join the Department of Zoology at the University of Chicago as an assistant professor. At Chicago, Davenport became increasingly interested in "quantitative biology." This idea, which became a guiding concept in Cold Spring Harbor science, was laid out in Davenport's 1899 *Statistical Methods with Special Reference to Biological Variation*. The book established Davenport's reputation as a leader in the effort to make biology less descriptive and more quantitative, like physics and chemistry.

Davenport took the empirical, experimental style of science he had learned at Harvard and Chicago out of the laboratory and into the salt marsh. His summer course in Field Zoology at Cold Spring Harbor became very popular, perhaps partly because much time was spent out of doors in bathing suits and casual clothing! The lectures, too, were held out of doors, with much discussion arising spontaneously as examples of the abundant marine life were fished from the harbor and their life histories, anatomy, and evolution were recounted with quiet enthusiasm by Davenport or his well-liked assistant Herbert E. Walter. A product of this pursuit of firsthand biology was "The Animal Ecology of the Cold Spring Sand Spit, with Remarks on the Theory of Adaptation,"

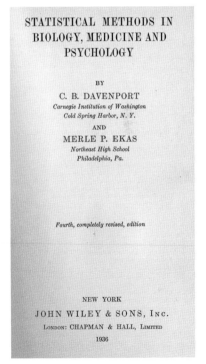

STATISTICAL METHODS IN BIOLOGY, MEDICINE AND PSYCHOLOGY

BY

C. B. DAVENPORT
*Carnegie Institution of Washington
Cold Spring Harbor, N. Y.*

AND

MERLE P. EKAS
*Northeast High School
Philadelphia, Pa.*

Fourth, completely revised, edition

NEW YORK
JOHN WILEY & SONS, Inc.
London: CHAPMAN & HALL, Limited
1936

Title page of Davenport's Statistical Methods.

Biological Laboratory course announcement.

his contribution to the decennial celebration of the young University of Chicago, in 1903. This paper was hailed as a landmark contribution to the new science of animal ecology.

Davenport began to turn the Laboratory into a serious scientific research and teaching institution, and by 1902 the Laboratory had grown in both quality and quantity. He quickly made a college degree and a serious interest in and capacity for original research prerequisites for entrance into the summer courses. New courses were offered, including Animal Bionomics, Ecology, and Evolution. Independent research was also now offered formally, made possible by the acquisition of new housing, so that spare rooms were now available to house investigators not enrolled in any course.

But even as the science at the Biological Laboratory became more advanced and professional, it began to run into financial trouble. Concerned that the Laboratory might outgrow the capabilities of the Brooklyn Institute to manage it, in 1902 the Wawepex Society considered ending its lease of the Laboratory grounds. Perhaps a university could better manage a rapidly growing and increasingly scholarly institution. Davenport, however, would have none of such a plan. He fired off letter after impassioned letter to Townsend Jones and the Wawepex Society, arguing for the desirability of maintaining the relationship with the Brooklyn Institute. The event that ultimately persuaded them, however, was Davenport's successful attempt to establish a major, year-round, scientific research institution at Cold Spring Harbor.

The Carnegie Institution of Washington

In 1902, industrialist Andrew Carnegie gave $10 million to endow the Carnegie Institution of Washington (CIW), a foundation to strengthen American education and research in science. The Institution would "in the broadest and most liberal manner encourage investigation, research and discovery [and] show the application of knowledge to the improvement of mankind." The areas for research were to be in astronomy, biology, and earth sciences. From the outset, the CIW supported science both intra- and extramurally, although opinion on which would be the most productive differed among the Carnegie's officers.

In March 1902, the Trustees appointed expert Advisory Committees to give them guidance on what fields of study were worthy of support. The first volume of the Carnegie Institution of Washington Year Book contained an appendix describing nine "Proposed Explorations and

Portrait of Andrew Carnegie, 1913.

Investigations on a Large Scale" of "great interest and importance" to the Carnegie but whose consideration had been deferred by the Advisory Committees. One of these proposals was for a "Biological Experiment Station for Studying Evolution" from Davenport writing from the University of Chicago.

---

280   CARNEGIE INSTITUTION

STANFORD UNIVERSITY, CAL., *August* 28, 1902.
To the Board of Trustees of the Carnegie Institution.

GENTLEMEN : I have received a copy of an outline of a plan for a proposed Vivarium for Experimental Evolution, suggested by Mr. Roswell H. Johnson, of the University of Wisconsin.

Permit me to say that, in my judgment, the establishment of such an institution, if placed in charge of some man of high skill in handling this class of experiments, would be an extremely desirable piece of work in the line of advanced research, and I trust that the Committee of the Carnegie Institution in charge of this matter will give it careful consideration.

Very truly yours,

DAVID S. JORDAN.

BIOLOGICAL EXPERIMENT STATION FOR STUDYING EVOLUTION

BY CHARLES B. DAVENPORT

ZOOLOGICAL LABORATORY,
UNIVERSITY OF CHICAGO, *May* 5, 1902.
To the Trustees of the Carnegie Institution.

GENTLEMEN : I beg leave to present for your consideration the following proposal for the establishment and maintenance by the Institution of a Biological Experiment Station for the study of evolution.

1. *Aims.* The aims of this establishment would be the analytic and experimental study of the causes of specific differentiation—of race change.

2. *Methods.* The methods of attacking the problem must be developed as a result of experience. At present the following seem most important :

(*a*) Cross breeding of animals and plants to find the laws of commingling of qualities. The study of the laws and limits of inheritance.

(*b*) The experimental production of variation, both by internal operations, such as hybridization, or by change of external conditions.

Page 280 of the Carnegie Year Book for 1902 showing Davenport's proposal for a "Biological Experiment Station for Studying Evolution."

Davenport's Campaign

Davenport had not been slow to grasp the opportunity offered by the possibility of Carnegie funding. Only 12 days after the Carnegie Institution was incorporated, he presented their board of directors with a plan to establish an endowed or fully supported laboratory that would be devoted to experimental studies of evolution. Presentation of the plan was Davenport's opening move in a campaign of 2 years during which he repeatedly recast and improved his proposal, visited other marine stations, lobbied friends and colleagues, prepared backup plans, and generally did everything in his power to establish a center for experimental biology at Cold Spring Harbor. His zeal is reflected in an August 18, 1902 letter to the CIW's John Shaw Billings, in which Davenport wrote

> I believe I am stating merely the cold truth when I say that in training and in work accomplished in the study of heredity and variation—the elements of evolution—no one in the country is as well prepared for experimental and quantitative studies in evolution as I am.…

To Davenport, the way to study evolution was, of course, quantitatively, by experiment, and not by passive observation. His new institution would use Mendelism and mathematics—especially probability theory—to study evolution and use it to benefit society.

The Carnegie Institution, however, had its eyes on Woods Hole, which was in serious financial straits. The Marine Biological Laboratory (MBL) trustees originally voted 60 to 3 to transfer the MBL lock, stock, and barrel to the Carnegie but there was a catch. The charter of the CIW directed the Carnegie trustees to establish *research* laboratories and they insisted the MBL discontinue its instructional program. After prolonged negotiations, and just when the deal seemed to have reached a conclusion, a majority of the trustees of the MBL, including Davenport, rejected the proposal.

Portrait of Francis Galton.

Portrait of Karl Pearson.

The demise of the plan for a Carnegie Department at Woods Hole offered a golden opportunity for Davenport, who revised his proposal, moderated the financial requests, and suggested that Cold Spring Harbor was the ideal site for a genetics research institution. Because Davenport already supervised an instructional program at the Biological Laboratory, he was happy to have the Carnegie Institution establish a facility dedicated to research. The Biological Laboratory would remain completely independent, although Davenport would direct both. The Wawepex Society agreed to lease a site to the Carnegie Institution for a nominal fee for a term of 100 years, renewable. On this site the Carnegie would build the new Station for Experimental Evolution. Davenport would resign his position at the University of Chicago and become the full-time director of the Carnegie Institution's Station at Cold Spring Harbor.

Davenport and his wife traveled to Europe and visited many marine biological laboratories in the late summer of 1902. They also arranged to visit the father of eugenics Francis Galton and his disciple Karl Pearson at the Galton Laboratory in London. Both were impressed by Davenport's enthusiasm but Galton declined to write a letter of recommendation, because of their short acquaintance, whereas Pearson wrote directly to the Carnegie Institution and commented, in respect to Davenport's personal qualifications for the post of director, that "Personally he seems to me stronger than his published work.... I should say he would not be wanting in energy and keenness of interest and would keep himself in touch with European workers and methods."

In the summer of 1903, Davenport made arrangements to spend a 3-month leave from the University of Chicago at the new Carnegie Desert Laboratory in Tucson, Arizona. The family was all packed to go when Davenport received the news. The Trustees had decided to establish a Department of Experimental Biology, consisting of a Station for Experimental Evolution at Cold Spring Harbor and a marine biological station at the Dry Tortugas, islands to the west of Key West. Davenport would receive $34,500 to establish the Station, a considerable amount of money in 1903. It is hardly surprising that Davenport wrote in his journal, "THIS IS A RED LETTER DAY!"

The Station for Experimental Evolution

Davenport resigned his position as associate professor at the University of Chicago in anticipation of his Carnegie appointment beginning on January 19, 1904. There was some discussion about the name of the Carnegie Station at Cold Spring Harbor. Billings suggested calling it simply the Biological Laboratory. Davenport disagreed strongly, offering instead the "Station for Experimental Evolution of the Carnegie Institution." Billings replied wryly, "I certainly do not think that the name should be 'Station for the Experimental Evolution of the Carnegie Institution.' While the Carnegie Institution may be in need of experimental evolution, I do not think that this institution is established for the purpose of evoluting it." The Department remained Experimental Biology, with Davenport's section entitled the Station for Experimental Evolution (SEE).

The formal opening of the new research laboratory took place on June 11, 1904. The Long Island Rail Road provided a special carriage for 50 guests from New York City. The principal speaker was none other than Hugo de Vries, one of the three who had "rediscovered" Mendel's work. de Vries's 1901 *Die Mutationstheorie* had sparked much discussion among biologists and was being hailed as one of the most important books since Darwin's *Origin of Species*. Davenport was justifiably thrilled to have one of the fathers of genetics as a guest speaker.

de Vries gave an address full of science and optimism. He hailed the establishment of a research institution to study evolution, for evolution "has to be attacked by direct experiment." The Station for Experimental Evolution was a lighthouse, a beacon

Portrait of Hugo de Vries.

Illuminated manuscript presented to Hugo de Vries on the occasion of the opening of the Station for Experimental Evolution and signed by guests and staff, including William Castle, George Shull, Frank Lutz, Anne Lutz, and Gertrude and Charles Davenport.

…illuminating the far more arid problems of the origin of species. It is surrounded by a denser darkness, for there is less previous knowledge in this field…the care of the lighthouse is given into the hands of Mr. Davenport and his staff, and many details of its internal affairs are looked after by the kind care of Mrs. Davenport. Thus provided, it can not fail to fulfill its mission, and to yield the results expected from it, and even more.

The new Carnegie Main Building was occupied on January 1, 1905. It had several small laboratories on two floors with a library on the second floor. The grounds, comprising approximately 10 acres leased from the Wawepex Society and adjacent to the Biological Laboratory, included the Victorian house facing the main highway. Since 1883 this

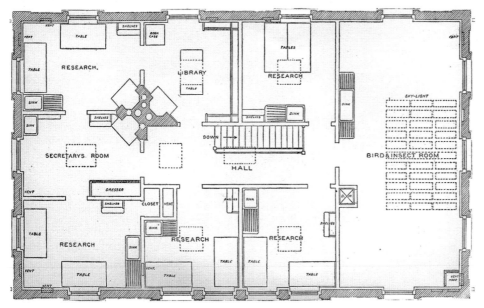

Carnegie building: (top left) *exterior,* (top right) *internal view of a laboratory, and* (bottom) *floor plan.*

house had been the residence of the Biological Laboratory's Director and his family; it was now the house of the director of both Cold Spring Harbor institutions. The grounds were rapidly furnished with pens, coops, and paddocks for the experimental menagerie, for at this stage all manner of organisms were subjected to Mendelian analysis.

Davenport's initial staff selections were drawn heavily from the graduate students at the University of Chicago, two of them already Cold Spring Harbor regulars. Maize geneticist George H. Shull and entomologist Frank E. Lutz had come out to Cold Spring Harbor for several summers during graduate school as instructors and researchers. There was also Anne Lutz, apparently no relation to Frank, who acted as Secretary of the SEE and also carried out important research. She was to identify suitable materials for hybridization experiments. Davenport's own work focused on the heredity of domestic races of birds, especially fowl. The addition of this new staff had the benefit of increasing the apparent size of the Biological Laboratory where they were cross-listed as visiting researchers in its annual report.

Portrait of George H. Shull.

Davenport enhanced this core of permanent staff scientists by appointing a distinguished group of biologists as Associates who were to devote their summers, or be part-time investigators during the school year, to research at the Carnegie Station. These Associates helped create an impression of size and prestige disproportionate to the Carnegie Department's fledgling status.

The Associates lent not just their names to the SEE. The scientific reports from SEE, most frequently printed in Carnegie Institution of Washington publications, were understandably sparse during the first few years. Although many of the SEE associates received research grants from the Carnegie Institution, they did not actually perform the research at Cold Spring Harbor. This did not deter Davenport from bulking up the list of SEE publications by including those of the SEE's Associates!

The Research Program

The science of the Carnegie Station in its early years ranged over the wide plain of agricultural genetics, animal breeding, and other questions of heredity. This was typical of the early days of Mendelian genetics when its application was being tested on all manner of organisms. In his 1905 report to the Carnegie Institution of Washington, Davenport listed no fewer than nine topics occupying his staff:

1. Investigations into inheritance and variability of plants.
2. Investigations into inheritance and variability of insects.
3. Investigations into inheritance and variability of other invertebrates.
4. Investigations upon aquatic vertebrates.
5. Studies on inheritance in domesticated animals.
6. Investigations into the cytological basis of heredity.
7. Cooperation with other investigators.
8. Work of subsidiary departments.
9. Care and development of the plant.

However, this list did not reflect a carefully thought-out plan but was rather a scattershot approach to studies of evolution and heredity. The program was, as MacDougal wrote, "as ill-defined as it had always been, and as it was to remain.... Varying lists of experiments had been proposed, but the differences in the successive lists did not represent progressive critical thought." Nevertheless, Davenport now had a building and a staff, and even without a clearly defined research program, the staff left to themselves found interesting problems.

A First Success: George Shull and Hybrid Corn

One of the main areas of investigation was agricultural genetics, headed in the first years by George Harrison Shull. While still a graduate student at the University of Chicago and doing a statistical study of variation of the floral parts of wild asters, Shull became deeply involved in work with the biometrical methods being developed by Davenport. He studied variation in many plants, and, by 1906, he was simultaneously conducting breeding experiments with no less than 45 different species of plants including Indian corn, the bean, the pink, foxglove (*Digitalis*), sunflower, tomato, poppy, potato, and tobacco. He was particularly interested in quantitative variations in hereditary characters in the evening primrose (*Oenothera*) and the shepherd's purse (*Capsella bursa-pastoris*). He produced no fewer than 22 papers on the former and 15 on the latter. Shull began to focus his interests on maize, growing his experimental work corn on a small plot of ground adjacent to the Carnegie station building. Here he first analyzed the inheritance of quantitative characters in maize, especially the number of rows of kernels per ear. By self-pollinating the corn he derived a number of inbred strains that differed in the average number of rows of grains on each ear. He observed that these lines declined in vigor and productivity as they became more uniform through inbreeding. When he crossed any two pure lines, however, the progeny were vigorous and highly productive, superior to either the parental inbred lines or the original open-pollinated corn. Shull's papers of 1908 and 1909 reporting on

George Shull's cornfield.

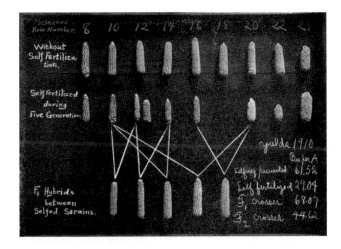

A display demonstrating hybrid vigor. The cobs in the first row are from plants grown in the field. The cobs in the second row are from plants that have been self-fertilized for five generations. These inbred lines are crossed and the ears from these hybrid plants are shown in the bottom row. The hybrid plants produce ears that are larger than both the original and inbred lines.

these experiments established the concept of "hybrid vigor" and became classics in agricultural genetics.

Between 1910 and 1913, Shull strengthened his early demonstration of the injurious effects of inbreeding in corn. He showed that cross-fertilization within the same pure line has little or no effect. It is only when different pure lines are intercrossed that hybrid vigor appears. The second generation of progeny within a pure line yielded 45 bushels of corn per acre; Shull's hybrids between pure lines yielded 68 bushels per acre.

Shull's work paved the way for the hybrid corn industry, especially after D.F. Jones at the Connecticut Agricultural Experiment Station in New Haven devised the "double-cross" method in 1917, using four different pure lines, crossing them two by two, and then intercrossing the two resulting hybrids. However, the new, higher-yielding varieties could not propagate themselves; the hybrid lines gradually diminished in yield. To maintain the high yield, farmers had to purchase hybrid corn seeds anew each year, creating a boon for the seed corn industry.

Through all this, the most important work of his career, Shull was seriously distracted from his own research. In 1905 the Carnegie Institution had made a large grant to Luther Burbank, the famous California plant breeder. Burbank had no high regard for Mendelian theory, instead carrying out mass hybridizations between plants collected in the wild, selecting those hybrid seedlings that looked promising to his eye, and inbreeding or self-fertilizing the most promising. Carnegie insisted that a well-trained plant geneticist would review and prepare Burbank's results for publication. Shull was this geneticist, and beginning in 1906, he spent some time annually in Santa Rosa working with Burbank.

It was a disaster. Shull quickly decided that Burbank's scientific methods were poor and his records worse than inadequate, while Burbank was not one to be told what to do by some young whippersnapper. Shull wrote a remarkably frank and well-balanced report for the Carnegie authorities, flatly stating the difficulties in carrying out the plan. Nevertheless, he was urged to keep at it, but by 1910 Burbank had stopped cooperating with Shull who complained bitterly about the interference of the Burbank project with his own experimental work. In 1913, it came to an end. Shull's report was never published.

Primroses and Fruit Flies

Anne Lutz.

While Shull was shuttling between California and Cold Spring Harbor, research on the genetics of plants was in the hands of Anne Lutz. Lutz had trained as a cytologist and carried out important cytological studies and hybridization experiments in plants. Her most notable finding was that Hugo de Vries's *gigas* mutant of *Oenothera* had double the number of chromosomes per cell as compared with the wild-type plant and was not a gene mutation of the kind that Morgan was studying in *Drosophila*. This condition, called polyploidy, gave *Oenothera* traits non-Mendelian patterns inheritance, which partly accounted for the strange behavior on which de Vries based his *Mutationstheorie*. The importance of Lutz's work was recognized at the time; the eminent cytologist E.G. Conklin called this an "epoch-making" result because it showed that mutations could be due to differences in chromosome number as well as intra-chromosomal mutations. Most unfortunately, Lutz and Davenport fell out, ostensibly over her caution in publishing her findings, but the situation was complicated by personal tensions between them. In 1910, Davenport asked for her resignation and she left the SEE in February 1911. Lutz carried out research in Belgium before returning to the United States but her scientific career was effectively ended. In 1932, she was awarded an honorary Doctor of Science degree from Purdue University, the first woman so recognized.

Frank Lutz was beginning research on variation in *Drosophila ampelophila,* later renamed *melanogaster,* as his experimental animal. At first Lutz's experiments did not pay off. He spent a year looking for inherited characters influencing development, but the flies were so sensitive to temperature that heredity almost always seemed to be swamped by environment. But he persisted, and found wing vein mutants and, in 1908, a dwarf mutant appeared in his cultures. Lutz published his results in 1911 after leaving the SEE. The title of the publication, "Experiments

Frank Lutz.

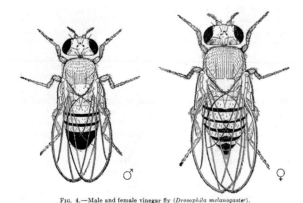

Fig. 4.—Male and female vinegar fly (*Drosophila melanogaster*).

Drosophila melanogaster.

with *Drosophila ampelophila,* Concerning Evolution" shows that he was true to the SEE's original intention, experimental evolution. The longest section was an analysis of more than 50,000 abnormal wings. Lutz made an attempt to interpret his findings in terms of Mendelian genetics but concluded that "… these abnormalities follow the spirit but not the letter of the Mendelian law." As Lutz's flies were confined to small vials and had little opportunity to fly, he also explored the effects of disuse on wing size. There appeared to be evidence of degeneration in flies of the 33rd to 35th generations, but Lutz made the mistake going one step further: flies of the 41st to 43rd generations had wings just as large as those of the 1st generation. In 1909, Lutz resigned his position at Cold Spring Harbor and moved to the American Museum of Natural History.

Although Lutz's own research may not have contributed much to the corpus of scientific knowledge, his contribution to *Drosophila* research was significant—he claimed that it was he who introduced T.H. Morgan to *Drosophila* as an experimental organism.

Davenport's Research Program

Davenport's early research had been on biometry, the measurement and statistical analysis of physical traits to differentiate natural groups of organisms. It might have been expected that his new institute would follow the same lines, but following the rediscovery of Mendel's work, Davenport rapidly became a fervent advocate for Mendelian genetics. His research at the SEE focused on examining the validity of Mendel's work across a wide range of organisms, and the SEE bred large numbers of animals. The annual report for 1906–1907 refers to the breeding of poultry, canaries, cats, sheep, and goats, insects of several species, and many species of plants, not to mention the mice, rats, guinea pigs, and rabbits of CIW Associate William Castle, visiting from Harvard

View of grounds with sheep pens.

University. In the period between 1904 and 1910, Davenport published papers on coat color in mice, the origin of black sheep, heredity of traits in poultry, the inheritance of plumage color and crest of canaries, transplantation of ovaries between hens, and, with Gertrude as coauthor, papers on the inheritance of eye color, hair color, and hair form in humans.

By 1912, Davenport's research was dominated by studies of heredity in humans. As we shall see in the next chapter, his enthusiasm for the practical applications of genetics to humans led him to become an enthusiastic proponent of eugenics.

The founding of the Carnegie Station for Experimental Evolution changed the intellectual tone at Cold Spring Harbor dramatically. The Carnegie Department set Cold Spring Harbor science moving in a new direction that would prove productive in generating much new variation of its own. This change was reflected in a change of name—the Station for Experimental Evolution became the Department for Experimental Evolution in 1906. In doing so, it conformed with the standard CIW practice of regarding its constituent units, no matter where located, as departments of the CIW.

For Davenport, the Department for Experimental Evolution marked the beginning of his scientific empire. For the next 30 years, Davenport's actions can be seen as calculated to advance science and benefit society in ways that would bring him recognition in the eyes of his peers and the general public.

GEORGE HARRISON SHULL DEMONSTRATES HETEROSIS IN CORN AND BESTS LUTHER BURBANK

Summary

Shull devised experiments that produced hybrid corn, providing a radically new method to improve crop yield and quality and ultimately leading to the "Green Revolution." He also assessed Luther Burbank's work for the Carnegie Institution and managed to debunk any scientific basis for Burbank's "fantastic" results.

The Discovery/The Research

The idea of hybrid vigor had been known for centuries, but its mechanism was a mystery. Charles Darwin in his 1876 work "The Effects of Cross and Self Fertilisation in the Vegetable Kingdom" provided many examples of deleterious inbreeding and beneficial crossbreeding. However, Mendel's laws, little known at the time, would be needed for interpretation of Darwin's examples.

Shull arrived at Cold Spring Harbor in 1904 before the Carnegie building was even finished and began his work on corn, inspired by the Danish botanist and geneticist Wilhelm Johannsen's work on inbreeding in beans. He wrote two critical, succinct, and precise papers describing his results. In "The Composition of a Field of Corn" published in 1908, Shull said that in an ordinary field of corn, individuals are complex hybrids. He noticed that all crossbred rows were similar in structure, vigor, and variability, but each self-fertilized row differed from the other self-fertilized rows. He determined that "[s]elf-fertilization soon eliminates the hybrid elements and reduces the strain to its elementary components." Regarding the cross-fertilized strains, "hybrids between nearly related forms are more vigorous than either parent." He concluded that the aim of the corn breeder should be to find and maintain best hybrid combination, not the best pure line.

The second paper, "A Pure Line Method of Corn Breeding" (1909), gave specific instructions for Shull's method. Breeders should self-pollinate the corn to produce a number of pure inbred lines and continue this inbreeding year after year. The resulting lines would decrease in vigor and productivity, eventually reaching the homozygous state. Finally, the inbred lines should be crossed to produce hybrids that are extremely uniform, highly vigorous, and productive.

These two papers were the basic method and insight into producing hybrid corn. Shull introduced the term "heterosis," short for "stimulation of heterozygosis," to describe this hybridization effect.

However, there was a problem with Shull's method — the inbred lines produced small amounts of seed and thus the method was expensive to carry out. This problem was solved by Edward M. East at the Connecticut Agricultural Experiment Station in New Haven, Connecticut and his colleague Donald F. Jones. They used a four-way (double-cross, two F1 hybrids crossed to produce seed) method that produced larger amounts of seed corn.

The idea of crossing inbred lines of corn was quickly taken up by agricultural experimental stations in the U.S. Midwest in the 1920s, and hybrid corn seed started to be produced on commercial scale. By 1939, the vast majority of U.S. corn planted was hybrid corn. In the 1960s, single-cross methods in which very closely related strains are used were developed.

Significance

Hybrid corn is one of the most significant agriculture advances of the 20th century. When hybrid corn was first cultivated, it immediately increased U.S. yields by 20%. Hybrid corn had better yield, was uniform in size for machine harvesting, and was drought resistant.

Continued

(Left to right) *George H. Shull, Arthur M. Banta, unidentified, J. Arthur Harris, Hugo de Vries (seated), A.F. Blakeslee, Charles B. Davenport, and Ross A. Gortner on the steps of the Carnegie Laboratory, 1912.*

The genetic basis of heterosis took much later to pin down. It was first thought that greater yield was due to suppression of deleterious recessives from one parent by the dominant alleles of the other. By the 1940s, the overdominance hypothesis held sway, in which the heterozygote is superior at key loci to either homozygote. This effect was later shown to be pseudodominance. In 1956 George F. Sprague proved showed additive and dominant effects were responsible.

The Scientist

Born April 15, 1874, on a farm near North Hampton, Clark County, Ohio, Shull had little early formal education. He decided to study botany at age 16, and from the age of 18 taught for next 5 years in Ohio public schools. After graduating Antioch College, he earned his Ph.D. in botany and zoology in 1904 at the University Chicago, where he met Charles Davenport, who was teaching biometry there. He remained at the Carnegie Station for Experimental Evolution at Cold Spring Harbor until 1915 before decamping to Princeton University, where he remained until 1942.

Shull was the founder and first editor of the journal *Genetics.* He was an active conservationist, playing a role in preserving the New Jersey Pine Barrens as a wildlife sanctuary under the National Park Service and participating in the fight to save New Jersey's Island Beach from developers. He served from 1928 to 1944 on the Princeton Borough Board of Education. He died September 28, 1954, in Princeton, New Jersey.

— J.C.

Cold Spring Harbor Becomes a Center for Eugenics, 1910–1940

Davenport Moves from "Basic" Human Genetics to Applied Human Genetics

Davenport developed an early interest in applications of Mendel's findings to human beings. He wrote in his 1909 report to the Carnegie Institution that although human genetics could not be experimental and hence fell outside the purview of the Station for Experimental Evolution strictly interpreted, nevertheless, "... the necessity of applying new knowledge to human affairs has been too evident to permit us to overlook it." Davenport began his research on human genetics with studies of eye color, skin pigmentation, hair color, and hair form, which were among the first applications of Mendelian genetics to human heredity. The research was done in collaboration with his wife, Gertrude Davenport, and she was first author on the four "Heredity of ... in Man" papers, which appeared between 1907 and 1910.

The paper on eye color ends with a discussion of the "... practical applications of these results to human marriage...," and Davenport quickly became more interested in these "practical applications" than in developing a science of human genetics. In 1909, he published a paper entitled "Influence of Heredity on Human Society" in which he argued, with fine rhetorical flourish, that "... the unrestrained, erotic characteristics of the degenerate classes are resulting in large families, which are withdrawn from the beneficent operation of natural selection by a misguided society that is nursing in her bosom the asp that may one day

Charles and Gertrude Davenport with their children, Millie (left), Charles, Jr., and Jane (right) (circa 1914).

WASHINGTON, D. C.
1910

AMERICAN BREEDERS MAGAZINE

Published by the American Breeders Association

A Theory of Mendelian Phenomena
PROF. W. J. SPILLMAN

New Hybridisation Method for Corn
DR. GEORGE HARRISON SHULL

Indian Cattle in the United States
A. P. BORDEN

Report of Eugenics Committee
DR. C. B. DAVENPORT

Breeding and Use of Tree Crops
PROF. J. RUSSELL SMITH

Hatching Quality of Eggs
DR. RAYMOND PEARL

Second Quarter · April, May, June

Vol. I · No. 2

Entered as second-class matter June 30, 1910, at the Post Office at Washington, D. C., under the Act of July 16, 1894.

American Breeders Magazine, *Issue 2 for 1910, includes two papers from Cold Spring Harbor: a report by Davenport on the activities of the Eugenics Committee and a paper by Shull on hybrid corn.*

fatally poison her." And he asked the question, "How shall the inroads of degeneracy be prevented and the best of our human qualities preserved and disseminated among all the people?"

The answer had been proposed by Francis Galton in late 19th century London. It was Galton who coined the term "eugenics," a movement dealing with "...what is termed in Greek, eugenes, namely, good in stock, hereditarily endowed with noble qualities." Galton uses the word "stock," and the American eugenics movement was intimately connected with the idea of improving "human stock."

The American Breeders Association had been established in 1903 with the purpose of promoting the use of Mendelian genetics for improving livestock and plants. Three years later, at the Association's second meeting, a Committee of Eugenics was set up, with Davenport as secretary, to investigate how human traits are inherited. To this end, subcommittees on Feeble-mindedness and Insanity were set up, followed by other subcommittees examining the heredity of criminality and pauperism; cancer; muscular strength; and intellectual traits. Another line of investigation would examine the characteristics of "normal" American families by carrying out surveys and by using the vast amounts of data already in the records of hospitals, prisons, and other institutions.

Davenport Establishes the Eugenics Records Office at Cold Spring Harbor

Davenport, as always eager to get on with the task at hand, conceived the idea of setting up an institute at Cold Spring Harbor for eugenics:

> One can not fail to wonder that, where tens of millions have been given to bolster up the weak and alleviate the suffering of the sick, no important means have been provided to enable us to learn how the stream of weak and susceptible protoplasm may be checked. Vastly more effective than ten million dollars to "charity" would be ten millions to eugenicists. He who,

by such a gift, should redeem mankind from vice, imbecility and suffering would be the world's wisest philanthropist.

He found the looked-for benefactor in his own neighborhood. Mrs. E.H. Harriman was the widow of Edwin H. Harriman, the railroad magnate who had owned, among others, the Illinois Central, Union Pacific, the Central Pacific, and the Southern Pacific railroads. Davenport persuaded Mrs. Harriman of the social good that would come from a scientific institution dedicated to eugenics, and she agreed to fund such an institution. Her patronage lasted nearly 8 years, and totaled nearly half a million dollars, with an additional $300,000 given for endowment. On February 16, 1910, Davenport wrote in his journal "A Red Letter Day for humanity!"

Edwin H. Harriman bought his first railroad at age 33 and went on to own many railroads.

The Eugenics Record Office Opens on October 1, 1910

Davenport described his aims and hopes in establishing eugenics as a scientific and social discipline in a lecture he gave in 1916, entitled "Eugenics as a Religion." Davenport offered a "creed" that included such statements as:

> I believe in striving to raise the human race, and more particularly our nation and community to the highest plane of social organization, of cooperative work and of effective endeavor.
>
> I believe that no merely palliative measures of treatment can ever take the place of good stock with *innate* excellence of physical and mental traits and moral control.
>
> I believe that … it is necessary to make careful marriage selection—not on the ground of the qualities of the individual, merely, but of his or her family traits….
>
> I believe in such a selection of immigrants as shall not tend to adulterate our national germ plasm with socially unfit traits.
>
> I believe in repressing my instincts when to follow them would injure the next generation.

Harry Laughlin and Charles Davenport outside the new Eugenics Record Office (ERO) building, circa 1913.

However, Davenport, now directing three institutes, must have realized rather quickly that he needed help and he found it in the shape of Harry Laughlin.

Laughlin was born in Oskaloosa, Iowa, in 1880. He began teaching agriculture at the North Missouri State Normal School (later Truman University) in 1900 and was an enthusiast for applying Mendelian principles to plant and animal breeding, particularly poultry. Davenport had published a monograph on *Inheritance in Poultry* in 1906, and Laughlin contacted him for information. Subsequently, Laughlin came to the Biological Laboratory in 1907 to take the genetics course, a visit that seems to have induced a sense of hero worship on Laughlin's part. Davenport, impressed by Laughlin's enthusiasm and interest, offered him the job of "superintendent," or manager, of the new Eugenics Record Office (ERO). For the next 25 years, Laughlin was a key figure in American eugenics.

The Trait Book *Catalogs Human Inherited Traits*

One of the goals of the ERO was to "… fill the need of a clearing-house for data concerning 'blood lines' and family traits in America." To do so required careful assessment of desirable and undesirable traits in families, and the first step was to define and catalog human traits. Davenport drew up a comprehensive list of human variants in *The Trait Book* (1912), which covered, it would seem, every conceivable human characteristic. The ERO's field workers used this when they went out into the world to interview families.

However, Davenport failed to recognize or ignored the fact that not all familial traits are biologically inherited. He listed some for which there was a clear genetic basis—hemophilia, Huntington's chorea, albinism—but also many for which there was little or no evidence of biological inheritance. For example, Davenport listed "hereditary traits" such

35	—CRIMINALITY
351	—CRIME AGAINST CHASITY
3511	—Adultery
3512	—Bigamy and polygamy
3513	—Crime against nature
3514	—Fornication
3515	—Incest
3516	—Prostitution
3517	—Seduction
352	—CRIME AGAINST PUBLIC POLICY
3521	—Counterfeiting
3522	—Disorderly conduct
3523	—Drunkenness
3524	—Incorrigibility
3525	—Perjury
3526	—Truancy
3527	—Violating liquor laws
3528	—Violating U. S. laws
3529	—Vagrancy
353	—CRIME AGAINST THE PERSON
3531	—Assault
3532	—Homicide
3533	—Rape
3534	—Robbery

Part of The Trait Book *showing the first entries under "Criminality."*

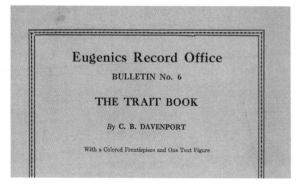

Cover of The Trait Book, *by Charles B. Davenport (1912).*

Harry Laughlin (**standing**), *teaching the Class of 1918 Eugenics Record Office field workers.*

as perjury, counterfeiting, and vagrancy under "Criminality—Crime against Public Policy."

Questionnaires and forms were sent on request and in mass mailing to families across the country and brought in person by field workers. The field workers were volunteers and were predominantly young women, many of them student nurses from the New York City area. New field workers spent a summer in training at the ERO and then took their "trait book" out into the field for a year, where they cataloged and documented the inheritance patterns of traits. The data were recorded on 3" × 5" cards and stored in a fireproof vault. By 1924, 750,000 cards had been filled out and filed. The pedigrees and family histories were analyzed for their fit to a simple Mendelian explanation.

In addition, the ERO sponsored studies on specific inherited characteristics, racial interbreeding, and studies of degeneracy in families and in institutions. Davenport, for example, published monographs on *The Hill Folk: Report on a Rural Community of Hereditary*

Eugenics Record Office—archives room with card index on far wall and field worker files on right.

Defectives (1912) and *Body-Build: Its Development and Inheritance* (1924). Davenport proposed that there was a hereditary trait, "thalassophilia" or love of the sea, to account for his finding that the families of naval officers often included a disproportionate number of seafarers (1919).

E. Carleton MacDowell, who came to Cold Spring Harbor in 1910 as a graduate student and returned in 1914 as an investigator, described the naïve enthusiasm of the eugenics students, running their calculating machines in the Carnegie building, tabulating data, and singing:

> We are Eu-ge-nists so gay
> And we have no time for play
> Serious we have to be
> Working for posterity

"Pedigree of Musical Capacity," Eugenics Record Office form including instructions to test sense of pitch, intensity, time, consonance, tone, and rhythm (top), and a family tree showing talent in vocal music and violin and piano (bottom).

Ta-ra-ra-ra-boom-de-ay,
We're so happy, we're so gay,
We've been working all the day,
That's the way Eu-gen-ists play
Trips we have in plenty too,
Where no merriment is due,
We inspect with might and main,
Habitats of the insane.
Statisticians too are we,

In the house of Carnegie,
If to future good you list,
You must be a Eu-gen-ist.

Laughlin, the ERO, and Social Policy: Sterilization and Immigration

Harry Laughlin turned his eugenic zeal increasingly toward politics and social policy in two particular areas—sterilization and immigration. Laughlin wrote in 1914 that

[t]here are two phases of the practical application of the Eugenics program. The first is concerned in fit and fertile matings among the upper levels. The second is concerned in cutting off the supply of defectives.… .To purify the breeding stock of the race at all costs is the slogan of eugenics … [with] … the ultimate end of practically cutting off the great mass of defectiveness now endangering the conservation of our best human stock, and consequently menacing our national efficiency and happiness.

In 1922, Laughlin published *Eugenical Sterilization in the United States*, in which he exhaustively analyzed the legal standing of sterilization in each state and gave extended case reports. As of January 1, 1921, he wrote that 3233 sterilizations had been performed, of which 2558 were in California. Laughlin devoted 20 pages of the book to his model legislation for sterilization laws.

There was a long history of concerns about immigrants polluting the American germline, the first immigration laws having been passed in 1882. There was a huge increase in immigration to the United States in the early decades of the 20th century and by the 1920s, eugenics arguments became important factors in bolstering political advocacy for immigration control. Davenport presented a motion at a meeting of the Committee on Immigration of the Eugenics Research Association in February 1920. It began "Whereas the protection of the germ-plasm of

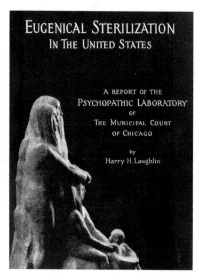

Cover of Laughlin's 1922 book, **Eugenical Sterilization in the United States.**

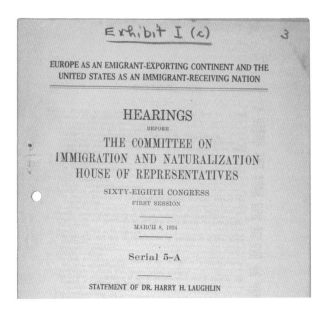

Cover of one of Laughlin's reports to the Committee on Immigration and Naturalization.

the nation is of prime importance for the future of the United States as a place in which to live and bring up families."

In 1921, Albert Johnson, chairman of the House Committee on Immigration and Naturalization, appointed Harry Laughlin as the Committee's "expert eugenics agent." Laughlin's fact-finding surveys, extensive tables and charts, and presentations before the Committee were persuasive and the 68th Congress passed the Immigration Act of 1942. This severely restricted immigration from countries deemed to have higher percentages of defectives, notably Eastern and Southern Europe and the Slavic area including the Balkans.

The ERO Becomes Part of the Carnegie Institution of Washington

In 1920, the ERO was assimilated by the Carnegie Institution of Washington, but this had little effect on the eugenic activities of Davenport and Laughlin. The 1922 annual report gives a flavor of the range of this work.

Arthur Estabrook specialized in examining "cacogenic" families—that is, families in which there was evidence of the inheritance of degenerate qualities. Dugdale's *The Jukes: A Study in Crime, Pauperism, Disease and Heredity* (1877) was the exemplar. Estabrook reported that he had completed his studies of a degenerate family in Indianapolis, the Tribe of Ishmael, first described by the Reverend Oscar C. McCulloch in 1888. There were, Estabrook wrote, "… three outstanding characteristics of the members of the Tribe: pauperism, licentiousness and gypsying." Their "… germ plasms have now spread through the whole middlewest and are continuing to spread the anti-social traits of their germ plasm with no check by society."

Laughlin's work appeared under the heading "Genetic Constitution of the American People." He described his work on immigration and declared that an effective policy on deportation was necessary to prevent "… contamination of future American stocks by the permanent introduction of excessive amount of defective alien germ-plasm."

The Demise of the ERO

By the 1920s, the intellectual and emotional climate in which eugenics had flourished began to change. Prominent biologists such as T.H. Morgan and H.S. Jennings began to speak out against eugenic policies, and Raymond Pearl described the literature on eugenics as largely

> … a mingled mess of ill-grounded and uncritical sociology, economics, anthropology and politics, full of emotional appeals to class and race prejudices, solemnly put forth as science, and unfortunately accepted as such by the general public.

The ERO publication, *Eugenical News,* was an embarrassment, especially when Laughlin, having analyzed the new Constitution of the German Republic, wrote of his admiration for its many provisions "for the maintenance of racial vigor and fecundity of the German stock."

Not surprisingly, the Carnegie Institution of Washington became concerned about the growing ill-reputation of the ERO, and how that reflected on the CIW itself. In 1929, the CIW president John C. Merriam assembled a special committee of scientists to assess the activities of the ERO. The committee criticized the ERO for being, of all things, too qualitative in its analysis of human traits. This must have been a blow to Davenport, the arch-exponent of quantitative biology!

In 1934 Davenport reached 65, the mandatory age of retirement for the Carnegie Institution of Washington. He stayed on as secretary of the Carnegie Department, as he had done at the Biological Laboratory since his retirement there. Albert Blakeslee, who had been at the Department of Genetics since 1916 and had become a member of the National Academy of Sciences in 1929, was named Acting Director in 1935 and became Director in 1936.

Without Davenport's support, Laughlin and the ERO were vulnerable, and perhaps sensing a window of opportunity, in 1935 the Carnegie Institution formed another investigatory committee. This committee concluded that the ERO record collection was useless for the study of human genetics—a waste of time, space, and, most importantly, money. The traits analyzed were in many cases too speculative and subjective to be of value, and the records, ironically, were so numerous that much of the data sat unanalyzed. The ERO should "cease from engaging in all forms of propaganda and the urging or sponsoring of programs for social reform or race betterment such as sterilization, birth control, inculcation of race or national consciousness, restriction of immigration etc."

One might have thought that Laughlin would have taken the admonishments of the committee to heart, but he continued much as before, pursuing his political activities with vigor. He had abandoned all attempts at concealing his political agenda and had lost touch with any good science being done in human genetics. He wrote with approval about the sterilization laws in Germany, and in 1936 he proudly accepted an honorary degree by the Nazi-controlled University of Heidelberg.

In 1938, Merriam stepped down as president of the CIW and was succeeded by Vannevar Bush. Bush wasted no time asking for Laughlin's resignation, ostensibly over concerns for Laughlin's deteriorating health (ironically for a eugenicist, he had had epileptic seizures, once while driving his car).

At the end of 1939, the ERO shut down permanently, and Laughlin returned to Missouri, where he died in 1943. Davenport wrote Laughlin's obituary for *Science* and gave his final verdict of the importance of Laughlin's work:

> One can not but feel that a generation or two hence Laughlin's work, in helping bring about restricted immigration and thus the preservation of our country from the clash of opposing ideals and instincts found in the more diverse racial or geographical groups, will be the more widely appreciated as our population tends toward greater homogeneity.

Eugenics had a relatively long life at Cold Spring Harbor. It arose in the social and cultural environment of the 1910s and 1920s, and although some scientists opposed it from the start, many, including some very eminent scientists, supported it. It appealed to those who had a desire to improve the human condition and a naïve faith in their understanding of nature and in the power of science to solve social problems. For all the hard work of the field workers, the enthusiasm of families mailing in their forms, and the hundreds of thousands of index cards, very little of scientific value came from the ERO.

DAVENPORT'S MISGUIDED QUEST TO IMPROVE HUMANS

Summary

After doing breeding experiments with a wide range of organisms and establishing the laboratory as a breeding station, Davenport settled on applying Mendelism to humans. This resulted in the publication of some influential books and the establishment of the Eugenics Record Office. The negative eugenics propounded by Davenport, although a reflection of the times, was sloppy science influenced by social prejudices and finally was discredited.

The Discovery/The Research

Davenport considered himself Francis Galton's scientific heir and very early he had introduced statistical methods to the study of evolution. His aim was to provide definitive proof of Darwinian natural selection using humans.

Davenport's first foray into this area was a series of papers on eye, skin, and hair color published with his wife Gertrude Crotty Davenport, who had a degree in zoology from "the Annex," later Radcliffe College. In their 1907 paper

Davenport working at his desk in the Eugenics Record Office.

"Heredity of Eye-Color in Man," they asked, "Is human eye-color inherited in Mendelian fashion?" They did not collect the data on 132 people for this paper themselves, but relied on "school principals and other friends." Blue, brown, gray, and black eye colors were considered, and the Davenports concluded that blue eye color was recessive to brown. Modern genetics studies have improved on this simplistic view and have shown eye color to be a multigenic trait.

"Heredity of Hair Color in Man," published in 1909 by both Davenports, studied factors concerning hair color in humans.

After 1910, with the establishment of the Eugenics Record Office (described in the main text), Charles Davenport was too busy as an administrator to do serious scientific work and he turned to crusading for the study and application of eugenics. Davenport did not waste any time in putting out his message and published *Eugenics: The Science of Human Improvement by Better Breeding* in 1910. While Davenport had obtained some reliable pedigrees for hereditary ataxia, Huntington's chorea, and color blindness, he also collected more spurious pedigrees for behavioral traits like business genes, warrior genes, and love of the sea. In addition, he tried to show the Mendelian nature of "traits" such as feeblemindedness, moral control, and pauperism.

Davenport believed that everything that "ran in families" was hereditary, and he did not take into consideration the effects and influences of, for example, environment, health, and nutrition.

Although many of his studies were eventually discredited, Davenport's enthusiasm for eugenics had some serious consequences. Aiding Davenport in his push for eugenical solutions was Harry Laughlin, who became superintendent of the Eugenics Record Office in 1910. Laughlin's testimony before Congress helped passage of the Immigration

Continued

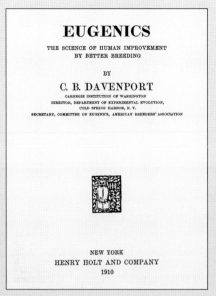

Title page of Eugenics: The Science of Human Improvement by Better Breeding.

Act of 1924 (The Johnson–Reed Act), which tightened immigration quotas and barred Asian immigration to the United States. At the state level, Laughlin participated in drafting laws for forced sterilization of people regarded as "unfit," and his "model" eugenical sterilization law was influential in the development of the 1933 Nazi Nuremberg racial code.

The Scientist

Davenport was born June 1, 1866, on the family's Davenport Ridge farm, near Stamford, Connecticut. His mother encouraged his interest in biology, but his strictly religious father thought he should pursue a more practical career— thus, his early study of engineering. Throughout his life, Davenport was a whirlwind of activity, being on the editorial boards of eight journals, a member of 64 societies, and the author of some 436 papers. He was said to begin work at 6 a.m. and was an enthusiastic walker, traveling as much as 15 miles by foot to some of his appointments. Davenport died February 18, 1944 in Huntington, New York. Ever the scientist (and perhaps egotist), Davenport willed his body to science and asked that "his brain be preserved in formalin."

— J.C.

4

Davenport Continues an Ambitious Research Program at the Station for Experimental Evolution

What of Davenport's other charges—the Biological Laboratory and the Station for Experimental Evolution—while he was engaged in promoting his eugenics initiative? His program of research between 1910 and 1920 might at best be described as eclectic, some elements remaining from the original goals of the SEE and others at the forefront of contemporary experimental biology.

Experimental Evolution Continues to Be a Focus of Research

In the progressive early years of the 20th century, Davenport asked: How can the processes of evolution be brought under controlled circumstances amenable to observation and dissection? How can science be put to human good?

Goose Island is an example of one of the less successful experiments in evolution. In 1909, at Davenport's urging, the Carnegie Institution purchased a small island in Long Island Sound, off the coast of Connecticut, where, Davenport argued, Carnegie researchers could study evolutionary processes in nature. In particular, Davenport intended to investigate the changes that a domesticated species undergoes in becoming feral by introducing white rats to the island and determining whether in an isolated population there would be a reversion to the wild-type brown coat color. The experiment began in 1911 in

Photographs of Goose Island.

collaboration with Philadelphia's Wistar Institute and immediately ran into difficulties.

The effort to kill off the native brown rats turned out to be unsuccessful, and the tougher brown rats were seen fighting with and killing the introduced white rats. As Davenport reported of one visit to the island, "Two female albinos with marked ears were found dead from severe wounds, as though bitten to death by rats," and one year later no albino rats were trapped. The project became an inadvertent test of the relative "fitness" of brown and white rats. By the mid-1910s, Goose Island was sold, although as late as 1922, Albert Blakeslee, whose work will described later in this chapter, planted triploid plants on Goose Island to determine how they fared in competition against native diploid plants.

Goose Island was a grand idea that failed because its planners underestimated the complexity of natural ecosystems. An alternative strategy, followed by Arthur Banta, was to create an artificial environment that was under the control of the experimenter.

Banta, hired in 1910, set out to discover whether species traits could be lost through disuse. He had published a comprehensive study of the fauna of Mayfield's Cave near Bloomington, Indiana, and so his experimental question was clear: Would animals raised in the dark for generations eventually lose pigmentation or other hereditary traits? In 1910, Banta built an artificial cave on the Biological Laboratory grounds. A substantial structure, it was made of reinforced concrete and was 41 feet long, 8 feet wide, and 6.5 feet high. It contained three rooms equipped with tanks for fish and crustacea and shelves for containers for land animals. The experiment began with a veritable zoo of species in the cave, including hydras, spiders, snails, copepods, millipedes, fruit flies, sowbugs, guinea pigs, sunfish, goldfish, crayfish, and frogs. Banta also tried the converse experiment of keeping the normally unpigmented cave forms of certain species living and developing in the light.

In one extended series of experiments, Banta raised tadpoles of the tiger salamander (*Ambystoma tigrinum*) in total darkness. The normally dark animals grew up pale when raised in the cave. A heritable, environmentally determined conversion from dark to light skin color would have been of obvious interest, but Banta could not say whether or not such an effect was heritable. Year in, year out, Banta continued to visit natural caves to collect animals and wrote reports of the cave experiment for the CIW annual handbook. These reports became shorter and shorter until they vanished without a trace in 1921.

Experiments Begin on the Germline Effects of Alcohol

Banta was not the only scientist at the SEE investigating the extent to which environment could produce heritable changes in organisms. During these years, when the temperance movement was heating up, a debate raged over the eugenic implications of drinking. Did it weaken the genetic stock? Or did it rather kill off the weakest stock, allowing only the strongest to survive? G.C. Bassett came to Cold Spring Harbor as a guest investigator in July 1913 to examine the effects of alcohol

E. Carleton MacDowell.

on albino rats, assessing their reproductive success and the quality of the young, and whether the alcohol influenced the "germ plasm" (i.e., whether the effects of alcohol were hereditary). A year later, when Bassett moved to the University of Pittsburgh, his experiments were taken over by E. Carleton MacDowell.

MacDowell followed a technique developed by Charles Stockard of Cornell Medical College in New York, in which animals were exposed to alcohol vapor each day until they were in a deep stupor. Reduced fertility appeared in MacDowell's lines of rats, and, surprisingly, the second-generation descendants showed even greater deficits. This suggested that the effect was transmitted through the germ cells to subsequent generations. Germline transmission, however, does not necessarily mean that mutations in the genes had been induced, and MacDowell did not distinguish between genetic and cytoplasmic inheritance. Although the small sample prevented his drawing any firm conclusions, MacDowell decided that his results generally confirmed Stockard's Lamarckian conclusions that alcohol affected the germ cells, reducing the capacity for full mental development for at least two generations, even in the absence of further alcohol treatment.

In 1918, MacDowell was called overseas to serve in the war effort. In his absence, his assistant, Emilia Vicari, carried on his experiments. But by the time he returned in 1919, not only was the war over, but Prohibition had been enacted. Drinking dropped below the radar of publicly discussed (and fundable) social problems. MacDowell returned from the war to find his rats gone and all support for study of alcohol on the germ plasm dried up.

Biometry was still part of the SEE program. J. Arthur Harris had joined the SEE in 1907 and pursued a bewildering variety of projects through the 1910s. He began carrying biometric analyses of wild plants to determine whether traits in nature "… undergo progressive change in a definite direction." Two years later, he was examining fertility and fecundity in plants; the variation and inheritance of quantitative traits in garden peas; and plant teratology, the latter study requiring the analysis of 40,000 seedlings. Two years later, he had examined 500,000 seedlings. In 1914 he was examining quantitative variation in the juices of apples and pears differing in size and fertility, measuring the osmotic pressure and electrical conductivity of the juices. This work occupied him until his departure in 1924 without, it seems, having achieved anything significant.

J. Arthur Harris in 1924.

Oscar Riddle Begins Work on Hormones

An entirely different line of research was pursued by Oscar Riddle, who had first come to Cold Spring Harbor as a visiting investigator in 1910, becoming a research investigator in 1912.

It seems probable that Davenport invited Riddle to Cold Spring Harbor because of their shared debt to the University of Chicago's C.O. Whitman, although it was Riddle who was Whitman's principal disciple. Whitman had a large research program using pigeons to study the relationships of variations in plumage, sex determination, and physiology, but he placed little importance on Mendelism or the chromosomal theory of heredity. Following Whitman's unexpected death, the new

Oscar Riddle with a pigeon.

department chairman at Chicago quickly let it be known that Riddle had no permanent place in his department. Riddle found a temporary position in the Biochemistry department. He obtained a small grant to help maintain Whitman's pigeon colony and set about editing the mountain of notes and manuscripts Whitman had left, a massive project.

Davenport came to Riddle's rescue, offering him a job as a research associate in the Carnegie Department, funds to maintain the pigeon colony, and even promised to publish the Whitman papers as Carnegie publications. Late in 1913, Riddle and his pigeons moved to Cold Spring Harbor, and in 1919, three volumes of Whitman papers, the first two edited by Riddle, appeared as Carnegie publications.

Riddle continued to carry out biochemical and physiological studies on pigeons, becoming particularly interested in the physiology and biochemistry of sex development. In the 1920s this line of enquiry led him to endocrinology and what proved to be his major contribution to the SEE, although it was research increasingly at odds with the genetic work of his colleagues (page 123).

Drosophila *Genetics Comes to Cold Spring Harbor*

The increasing emphasis on genetics was exemplified by two appointments Davenport made in this period.

The first was Charles Metz in 1913. Although Frank Lutz had been an early pioneer in *Drosophila* research at Cold Spring Harbor, it was Metz who established *Drosophila* as a research program at Cold Spring Harbor. This was hardly surprising given Metz's training as a graduate student in T.H. Morgan's group at Columbia. It was only 3 years earlier that Morgan had observed a white-eyed male fly and found that eye color exhibited sex-linked inheritance. The short paper he published in *Science* changed genetics forever. With three undergraduates, Alfred Sturtevant, Calvin Bridges, and Hermann Muller, Morgan transformed his Columbia University laboratory into a "fly room," analyzing mutations in flies. Sturtevant made a major advance when he showed how the mutations could be mapped to specific points on the four *Drosophila* chromosomes, a technique that Bridges made his own (Chapter 8). Although the "gene" remained an abstraction, Morgan's chromosome theory gave the genes a physical place in the cell.

Charles W. Metz, 1922.

Metz accepted Davenport's offer even before finishing his Ph.D., perhaps because Davenport agreed to underwrite a field trip Metz wanted to make to collect *Drosophila* in the wild to study their chromosomal diversity. In January of 1914 he was on his way to Cuba. On his return, Metz examined the chromosome sets of some 30 species of *Drosophila* and several closely related genera, and found 10 or 11 types of chromosome complexes, all apparently derived from a single central, unspecialized type. Metz also showed that the familiar pattern of chromosome pairing during meiosis is true in *Drosophila*. The recognition that this pattern held true for *Drosophila* was an important generalization and provided further confirmation of the utility of the fruit fly as a genetic model system.

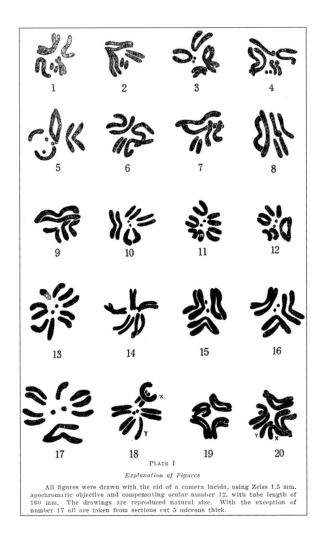

PLATE I

Explanation of Figures

All figures were drawn with the aid of a camera lucida, using Zeiss 1.5 mm. apochromatic objective and compensating ocular number 12, with tube length of 160 mm. The drawings are reproduced natural size. With the exception of number 17 all are taken from sections cut 5 microns thick.

Metz's camera lucida *drawings of the chromosomes of different* Drosophila *species.*

Plant Genetics Continues at Cold Spring Harbor

Davenport's second significant appointment was to replace George Shull who had left Cold Spring Harbor in 1915 with Albert F. Blakeslee. Blakeslee had been an assistant in botany at the Biological Laboratory during the summers of 1901 and 1902 while he was still a graduate

Albert F. Blakeslee.

student at Harvard. In 1912 he was on the faculty at the Connecticut Agricultural College when he accepted Davenport's offer to become a research investigator at Cold Spring Harbor.

Blakeslee's doctoral research had already made him well-known among naturalists for he discovered mating types (i.e., sexes) in the mold *Mucor.* But after moving to the SEE, he began a new line of research on the Jimson weed, *Datura stramonium.*

The work was a refinement of the kind of research that had been carried out by Anne Lutz on the shepherd's purse, *Capsella bursa-pastoris.* Lutz had shown that mutations could consist of wholesale duplications of chromosomes. What Blakeslee did was to associate changes in the numbers of specific chromosomes with changes in specific traits.

Although most plants and animals have two copies of each chromosome and are called diploid, there can be significant variation in chromosome number and composition. An organism with three copies of a chromosome is called trisomic for that chromosome. Trisomies occur in humans: The best known is trisomy 21, which results in Down syndrome.

Seed capsules of Datura *mutants with a schematic representation of the chromosomes of each.*

Blakeslee found a *Datura* plant with three copies of a chromosome, the extra chromosome causing a mutant seedpod that Blakeslee called "Globe." Blakeslee focused his research on *Datura* trisomies and other chromosomal changes, raising as many as 70,000 *Datura* plants each summer. By the time he retired in 1941, he had found a trisomic type for each of *Datura*'s 12 chromosomes, each trisomy with a recognizable phenotype of the seed capsule.

Blakeslee was joined by John Belling, a gifted cytogeneticist. Belling developed the technique of using acetocarmine to stain the chromosomes of plant cells smeared on a glass slide. The chromosomes could be observed in fresh tissue in a matter of minutes and the method was soon universally adopted—not only by plant cytologists but also by zoologists. It also produced a better microscopic view of the entire number of chromosomes in a cell's nucleus, as it avoided cutting the chromosomes into fragments. Using this technique, Belling and Blakeslee refined their analysis of *Datura* polyploidy, as well as determining the chromosome numbers for a great many species of flowering plants.

John Belling.

The Biological Laboratory Continues Its Course Program

At the Biological Laboratory, the program of courses continued essentially unchanged until 1917, when the course in Cryptogamic Botany, which had long been taught by Harlan H. York of Brown University, became a casualty as the professor entered war service. The number of students in all courses combined was little more than half what it had been in the previous years. In the following year, a wartime course in Sanitary Entomology was inserted. It was taught by Elizabeth H. Wright, who had taught a course in Economic Entomology in the previous summer. The course in Systematic and Field Botany was also listed as a "war course." The quality of the students remained high in the courses, however, as Davenport sought to attract increasing numbers of professional biologists and graduate students.

Clarence Little Brings Cancer Research to Cold Spring Harbor

Clarence Cook Little came to Cold Spring Harbor after finishing his Ph.D. with noted geneticist W.E. Castle at Harvard's Bussey Institute. He set out with tremendous, unfocused energy, studying inheritance in

Clarence Cook Little.

dogs, cats, pigeons, canaries, sheep, and humans. He studied the variations of the human sex ratio under the auspices of the ERO, and he carried out tests of the viability of the spermatozoa of various animal species in acid solutions of a range of concentrations. Collaborating with Little in a study of skull and body size in mice was Reginald G. Harris, an assistant who would later have considerable influence on the Laboratory.

It was in the genetics of cancer, however, that Little made his name. Working with Leonell C. Strong, a student at Columbia, Little examined the susceptibility of the Japanese waltzing mouse to a transplantable sarcoma. He found that 100% of the Japanese waltzing mice inoculated with the sarcoma developed tumors, but none of the control mouse strain did. They crossed the two strains and found that the F_1 hybrids were

---

SPECIAL ARTICLES

THE HEREDITY OF SUSCEPTIBILITY TO A
TRANSPLANTABLE SARCOMA (J. W. B.)
OF THE JAPANESE WALTZING
MOUSE

In 1916[1] the writer in collaboration with Tyzzer reported on the inheritance of susceptibility to a transplantable carcinoma (J. W. A.) of the Japanese waltzing mouse. This tumor grew in one hundred per cent. of the Japanese waltzing mice inoculated and in zero per cent. of the common non-waltzing mice. When these two races were crossed, the F_1 generation hybrids showed sixty-one out of sixty-two mice to be susceptible. In these mice growth was as rapid if not more so than in the Japanese waltzing mice themselves. The one exception may well have been due to faulty technique for a reinoculation test was not made.

Little's paper from 1920 where he determines that three to five factors are needed to account for the hereditary susceptibility of mice to a transplanted sarcoma.

even more susceptible to the sarcoma. However, F_2 hybrids were almost 100% *resistant* to the sarcoma. Little and Strong continued a genetic analysis carrying out backcrosses. The end result was that they were able to estimate that as few as four factors were involved, none of which were carried on the X chromosome.

In 1921, Little became Assistant Director of the Carnegie Department, but his tenure was short lived. In July of that year, he resigned, apparently after disagreements with Davenport, but not before he had developed the C57BL inbred strain of mice, still one of the most widely used mouse strains. Little went on to become the president of the University of Maine and, subsequently, of the University of Michigan. In 1929, Little founded the Roscoe B. Jackson Laboratory, in Bar Harbor, Maine, a research institution dedicated to the investigation of mouse genetics.

Davenport Attempts to Integrate His Three Institutions

A positive event for the Biological Laboratory came in 1917 when it was formally incorporated as a Department of the Brooklyn Institute of Arts and Sciences (BIAS). Occasional financial crises had occurred since 1902, when the Laboratory was nearly taken over by Columbia University. Making the Laboratory a proper department of the BIAS seemed to be a gesture of commitment and gave Davenport a sense of security.

Davenport did his best to integrate the activities of his three institutions, although those of the ERO were necessarily kept more distinct. Faculty were cross-listed in the annual reports, and the ERO was publicized in the Laboratory annual reports. Everyone, including the ERO staff, ate in the Laboratory's Blackford dining hall, which simultaneously united the three institutions into a close-knit community. Researchers at all three institutions offered evening lectures, open to visitors and staff of any of the three.

The Biological Laboratory curriculum leaned increasingly toward evolution and genetics, bringing it more in line with the activities of the ERO and the Carnegie Station. Nevertheless, teaching at the Biological Laboratory diminished in importance for Davenport as he focused on the prestige of professional science and on the political power and perceived social good of eugenics. Each year, more scientists visited the Biological Laboratory to pursue their own research. As an incentive to research, the Biological Laboratory began publishing the "Cold Spring Harbor Monographs," which collected the results of the summer investigators, and also offered scientists publication in the Brooklyn Institute's "Science Bulletin."

Even though Blakeslee, Belling, and Metz had brought genetics to the Carnegie Department, the latter still seemed more like a farm than a scientific laboratory: There were two large pigeon houses; sheep, poultry, dogs, and cats ranged the Carnegie grounds; the "cave" was still filled with subterranean creatures; and fields and greenhouses were filled with experimental crops. A baffling number of experimental organisms were matched by a remarkable range of topics, often being undertaken by the same researcher. Among the topics mentioned earlier, J. Arthur Harris compared the salt concentration of saps of tropical and desert plants; studied heredity in garden beans; collaborated with scientists at the Connecticut Agricultural Experiment Station to predict a hen's egg-laying potential by its pigmentation and other characters; pursued statistical investigations into the physiology of egg production and the variation in measurements on plot tests conducted by the Department of Agriculture; and even studied human metabolism.

The Department of Genetics Is Formed

An administrative change that must have been very gratifying to Davenport was a change in the name of the SEE. Typically, the various

The road sign for the Department of Genetics. As it retains the "Station for Experimental Evolution," the sign was probably made shortly after the change of name.

institutes supported by the CIW and scattered throughout the country were known as its Departments. For example, there was the Department of Embryology in Baltimore, Maryland, the Department of Terrestrial Magnetism in Washington, D.C., and the Department of Plant Biology in Stanford, California. The Station for Experimental Evolution, then, had always been something of an anomaly but in October 1920, "Experimental Evolution" disappeared from the title of the CIW's institute at Cold Spring Harbor, and it became the Department of Genetics. At the same time, as recounted in Chapter 3, the ERO was brought into the Department of Genetics and under the wing of the CIW.

It was increasingly clear that the Biological Laboratory was losing out to the Department of Genetics in the competition for Davenport's attention. Not long afterward, Davenport began to contemplate strategies for relieving himself of his duties there. The result was a dramatic change in direction for Cold Spring Harbor.

OSCAR RIDDLE DEVISES AN ASSAY IN PIGEONS THAT REVEALS THE ROLE OF PROLACTIN IN MAMMALS

Summary

Riddle's clever development of the pigeon crop sac assay enabled him to isolate prolactin, the hormone responsible for lactation in mammals.

The Discovery/The Research

In 1912, Charles Davenport invited Riddle to join the Station for Experimental Evolution at Cold Spring Harbor. They originally met when Davenport was teaching at the University and Riddle was Charles Otis Whitman's student studying the biochemical basis of dark and light bars on pigeon wings. When Riddle moved to Cold Spring Harbor in 1913, he brought the pigeons with him.

Heavily influenced by Whitman's thinking, Riddle emphasized metabolic and physiological explanations for biological processes. His initial research at Cold Spring Harbor was on pigment distribution and sexual determination and reversal in terms of biochemistry and physiology.

By 1928, it was known that extracts of the anterior pituitary induced lactation in rabbits, but reports at this time characterized only the growth-promoting and gonad-stimulating activities of anterior pituitary extracts. Riddle's proposal was that there was a third activity, prolactin, and he set about devising an assay to determine this. Riddle read colleague George Corner's report and decided to exploit the "crop milk" secreted by pigeons and doves (highly nutritious liquid of epithelial cells shed from crop lining and containing large amounts of protein and lipids), even though the crop gland is not homologous with the mammary gland. In his assay, Riddle injected anterior pituitary extracts and other hormones obtained from colleagues into pigeons to see if they could stimulate crop gland enlargement and result in the production of crop

"milk." The assay was very labor intensive and hundreds of birds had to be injected over the course of several days to produce a response. In 1932, Riddle found extracts of beef and sheep anterior pituitary that stimulated crop gland enlargement sevenfold over untreated birds. He identified the new hormone as prolactin and showed it induced lactation in male and female guinea pigs and female rabbits. He reported that milk secretion occurred after 2–3 days in rabbits and 3–5 days in guinea pigs and that "the crop-gland response is equally decisive in either species and sex; the gonad-stimulating response is more pronounced in males." Riddle received widespread press coverage for this discovery—he was portrayed on the cover of the January 9, 1939, issue of *Time* magazine.

Significance

Before the 1932 discovery of prolactin, it was believed that lactation was triggered by the placenta, corpus luteum, or ovaries. Human prolactin was not isolated until the early 1970s by Henry G. Friesen's lab in Canada and was made possible by the development of radioimmunoassay techniques. Prolactin has since been shown to be a critical hormone, having more than 300 different biological activities, including homeostasis, fluid balance, corpus luteum function in rodents, and metamorphosis in amphibians.

The Scientist

Riddle was born September 27, 1877, in Cincinnati (near Bloomington), Greene County, Indiana, to a farming family. He entered Indiana University in 1896, but then spent 2 years teaching in Puerto Rico and collecting specimens for U.S. government agencies. After his graduation he taught physiology at Central High School in St. Louis, Missouri

Continued

Oscar Riddle, 1942.

Throughout his life, Riddle was a strong proponent of the proper teaching of biology in public schools and fought against the influence of reactionary religious forces; in fact, his first published paper in 1906 was on the teaching of physiological zoology in secondary schools. He was the founder of the National Association of Biology Teachers. He married late in life at age 60 to Leona Lewis, whom he met on a cruise. The Riddles were known to entertain their dinner guests with shuffleboard played on a diagram painted on their dining room floor.

Colleagues described Riddle as being overconfident and dogmatic, and Riddle himself admitted "... my inexperience let me to an overconfidence and dogmatism which more mature years could only regret." Riddle died November 29, 1968 in Plant City, Florida.

— *J.C.*

before returning to the University of Chicago in 1906; he was awarded his Ph.D. in 1907.

ALBERT BLAKESLEE AND JOHN BELLING REVEAL CHROMOSOMAL TRAITS

Summary

In the early 1920s, Albert Blakeslee and John Belling showed that mutations arose not only from point mutations but also through duplications of entire chromosomes.

The Discovery

In 1915, Albert Blakeslee found a specimen of Jimson weed, *Datura stramonium*, with spherical seed capsules, a mutation he called *Globe*. Remarkably, this was not the only trait affected—the whole plant showed differences from the wild type. This was in marked contrast to the typical Mendelian mutation affecting only one trait—for example, changing *Drosophila* eye color from red to white. Blakeslee regarded this plant as a new species, arising according to de Vries' mutation theory. Blakeslee continued to search for similar mutants—at the height of his research program, Blakeslee was growing 70,000 plants each summer—and by 1920, he had found 12 such mutants, each clearly distinguishable in a number of traits.

In 1920, Blakeslee was joined by John Belling, a brilliant cytogeneticist. Belling had developed a much-improved method for making chromosomal preparations and staining them with his iron aceto-carmine stain. They set about determining the chromosomal basis of the 12 *Datura* mutants. They found that normal *Datura* plants had 12 pairs of chromosomes, for a total of 24 chromosomes, but that each mutant had 25 chromosomes—two each of 11 chromosomes but three of the 12th. The key observation was that the trisomic chromosome was different in each mutant, leading them to conclude that the characteristics of each mutant were caused by genes on the extra chromosome. Furthermore, they were able to relate the traits of the mutant plants to each of the extra chromosomes. They were able to show, for example, that *Globe* had three copies

An illustration from Blakeslee's paper showing the wild-type seed capsule (top) and the seed capsule for the trisomies of each of Datura's *12 chromosomes.*

of the S chromosome, whereas the mutant *Poinsettia* had three copies of the M chromosome.

Significance

Ann Lutz, one of the first investigators at the Station for Experimental Evolution (SEE), had shown in 1907 that *Oenothera gigas* was tetraploid, having twice the number of chromosomes as *Oenothera lamarkiana*. She went on to find other chromosomal abnormalities—for example, that

Continued

Balanced Types	Unbalanced Types		
Diploid	Modified Diploids		
(2n)	Simple Trisomic (2n+1)	Simple Tetrasomic (2n+2)	Double Trisomic (2n+1+1)
Triploid	Modified Triploids		
(3n)			
Tetraploid	Modified Tetraploids		
(4n)	Simple Pentasomic (4n+1)	Simple Hexasomic (4n+2)	Simple Trisomic (4n-1)

Blakeslee represents each of Datura's *12 chromosomes as a double bar (upper left). The remaining figures illustrate various chromosomal changes.*

Oenothera lata is triploid. But it was Blakeslee and Belling who extended the idea of mutation to include phenotypes arising from unusual numbers of chromosomes, even though these chromosomes were not carrying mutations of the kind studied by Morgan's group. Blakeslee also emphasized that the changes brought about in this way were an important mechanism in generating new species from old. He continued to work on chromosomal abnormalities and, with Charles Gager, in 1921 was able to produce mutants using radium to break up chromosomes. However, their paper was delayed until February 1927, and it was immediately overshadowed by a paper published in July, Hermann Muller's classic work producing mutations in *Drosophila* using X rays.

The Scientists

Albert Blakeslee came as a Staff Scientist in the SEE in 1915, taking the place of George Shull who had moved to Princeton University. Born in 1874, in Geneseo, New York, Blakeslee was Professor of Botany at Connecticut Agricultural College, Storrs before moving to Cold Spring Harbor. In 1934 he succeeded Davenport as director of the SEE, a post he held until his retirement in 1941. In addition to his work on *Datura*, Blakeslee became interested in human genetics, especially the inheritance of sensory traits such as smell and taste (see page 121).

Blakeslee's most important work was done with John Belling, a brilliant cytogeneticist. Belling was born in England in 1866 but left in 1902, first for Jamaica and then, in 1907, to the Florida Experimental Station. He came to Cold Spring Harbor in 1920 and there developed the aceto-carmine stain for chromosomes. He and Blakeslee published many papers together, but throughout his life Belling suffered from severe depression and was committed to an asylum for the mentally ill. In 1927 after three years in the asylum, Davenport arranged for Belling to go to the University of California, Berkeley for his health and continued to support him there. Belling died in 1933.

— J.A.W.

Belling and Blakeslee are seated in the middle *row, Blakeslee in shirtsleeves. Charles Davenport stands behind Blakeslee.*

5

The Long Island Biological Association Takes Over the Biological Laboratory

Following the war, enrollment in the summer courses at the Biological Laboratory dropped, and again the Brooklyn Institute considered dissociating itself from the Laboratory. The Carnegie Department remained well endowed and few believed the Laboratory could be in trouble, because many assumed it too had the Carnegie Institution's support. Davenport tried in vain to explain that although intellectually the two institutions were closely linked, financially they were completely distinct. "Those who might otherwise come to the support of the laboratory," he complained in a 1921 memorandum to the Laboratory trustees, "are led to believe that it is being taken care of by the Carnegie Institution of Washington, whereas it actually receives and can hope to receive nothing from it."

Through the impetus of Harold Fish of the University of Pittsburgh, who taught the introductory course at the Biological Laboratory, the University of Pittsburgh in 1920 contemplated acquiring the Laboratory. Davenport, disgusted with the situation and perhaps tired of the Biological Laboratory altogether, threatened to resign, but instead he began a campaign to raise local support for the Laboratory and to rescue it from being absorbed by a distant university.

First, he wrote in long memorandum of 1921 that there was a "crisis in the affairs of the Biological Laboratory." The Laboratory was being "outdistanced" by its long-time friendly rival on Cape Cod, the Marine Biological Laboratory at Woods Hole, as well as by its neighbor, the

Charles Davenport circa 1920.

MEMORANDUM CONCERNING BIOLOGICAL LABORATORY OF THE BROOKLYN

INSTITUTE OF ARTS AND SCIENCES.

I. The present is a crisis in the affairs of the Biological Laboratory. This crisis has been brought about by the following conditions.

A. Despite its strategic position, on a favorable part of the seafront; one hour from New York City with its scientific and medical population, its libraries and its stores of scientific supplies; adjacent to the large scientific establishment of the Carnegie Institution of Washington, the Biological Laboratory at Cold Spring Harbor has become quite outdistanced by that at Woods Hole.

B. An opportunity is offered us now to secure the best biophysicist in the country. If he should come to Cold Spring Harbor this laboratory would at once be placed in an outstanding position in the country.

C. New medical schools are arising in New York City and in the states to the westward which are looking for research opportunities at the seashore-irreplaceable for work in physiology, biochemistry and biophysics.

D. The University of Pittsburgh has under consideration the possibility of securing the plant and rights of the Biological Laboratory. The Laboratory may and should be lost to local control if local interests are not able to provide for its proper development.

E. The identity and special needs of the Biological Laboratory are being eclipsed by the growth of the department of Genetics of the Carnegie Institution of Washington and those who might otherwise come to the support of the laboratory are led to believe that it is being taken care of by the Carnegie Institution of Washington, whereas it actually receives and can hope to receive nothing from it.

F. The present director of the Laboratory must resign his position at once as the Carnegie Institution of Washington demands his whole time. He has already been given by the Institution seventeen years to put the affairs of the Laboratory in order and turn it over to another director. Increasing cares and growing years strengthen the demands of the Institution.

This crisis may be met in various ways.

1. The Brooklyn Institute may be advised to close the Laboratory as not worth the effort to maintain it. Against this may be urged:

a. The attendance at the Laboratory is growing and is considerably greater this season than ever before. This demonstrates that the Laboratory is fulfilling a needed service in instruction and research in biology.

b. The opportunity to do a great service that will advance knowledge and bring credit to the community is not lightly to be set aside.

c. The Institute has already trust funds ($27,500) to administer for the Laboratory.

Davenport's 1921 memo.

Carnegie Department of Genetics. Worse, it was in danger of being lost to the University of Pittsburgh. All this at a time when the rise of medical schools in the New York area made the experimental opportunities available at a seaside marine station "irreplaceable." Finally, Davenport wrote, he had no more energy to give the Laboratory; the Carnegie Department, he said, demanded his full-time attention. Davenport then laid out a plan to save the institution.

The Laboratory ought to form a local committee to discuss plans and raise money, and a scientific advisory board should be set up. The Laboratory ought to have a resident director to administer and maintain it throughout the year. The Laboratory ought to purchase the 40 acres surrounding their current 3 acres, for additional laboratories as needed and so that resident scientists could build houses near the Laboratory. Further modeling his plan on the successful one pursued at Woods Hole, the Laboratory ought to reinstate its biological supply department to provide biological materials to high schools and colleges. Woods Hole was able to make a nice little profit from its biological supply department (Davenport provided the figures); a similar operation would support the research programs at Cold Spring Harbor.

Davenport seems to have seen in biophysics a new opportunity for the Biological Laboratory to rejuvenate itself. This fledgling discipline resulted from a cross of physiology with new techniques for measuring and describing the properties of biological materials. He wanted to hire Professor William T. Bovie of Harvard Medical School—"the best biophysicist in the country"—to jump-start a program in biophysics. Bringing Bovie to Cold Spring Harbor was "of first importance," he wrote in the 1921 memo, "if the Laboratory is to be kept at the front of modern research and instruction." Although he was not explicit, Davenport seems to have intended that Bovie be a "resident" investigator (i.e., the first full-time researcher at the Laboratory).

The plan outlined a dramatic shift in the philosophy of the Biological Laboratory. There would be less emphasis on instructional summer courses of a popular nature and more on training advanced students in biological research. It would become a year-round research institute with a program in biophysics in a purpose-built laboratory, and there would be a major land acquisition. Of course, such an expansion would require new facilities and expensive instrumentation, and Davenport set about finding the necessary funds through a new entity, the Long Island Biological Association (LIBA).

The Long Island Biological Association

The LIBA was incorporated in 1923 with the goal of taking over the finances and administration of the Laboratory, beginning in 1924. The Brooklyn Institute transferred to LIBA ownership of Blackford Hall, the several dormitories, and the endowment and scholarship funds. The Wawepex Society turned over to the LIBA the Jones Lab and the lecture hall and drafted a 50-year lease for the grounds.

A first priority was to raise funds for the new association. Davenport took full advantage of the Laboratory's prime location among the gentry of Long Island's Gold Coast. The LIBA Board of Managers and roster

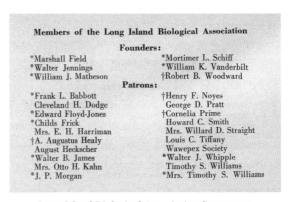

Long Island Biological Association Supporters.

William K. Vanderbilt (left), *in uniform on the yacht* Alva, *and Otto H. Kahn* (right), *supporters of the Long Island Biological Association.*

Otto Kahn's Gold Coast mansion, Oheka Castle.

of Founders, Patrons, and Sustaining Members bristled with some of the biggest names in American finance and industry: William Vanderbilt, Louis Tiffany, J.P. Morgan, Marshall Field, Mortimer Schiff, August Heckscher, Childs Frick, and of course Jones and Harriman. Prominent scientists, too, adorned the board and membership lists, including Herbert Spencer Jennings, T.H. Morgan, and Sewall Wright.

The first meeting of the LIBA was in February 1924. The 1925 Biological Laboratory Annual Report listed six Founders and 20 Patrons who had contributed at least $5000 and $500, respectively, and 170 Sustaining Members who made an annual contribution. There was also a Women's Auxiliary Board with responsibility for arranging a visiting day at the Laboratory and the formation of children's classes for nature study. Mrs. Otto Kahn was a member of the Women's Auxiliary; she and her husband lived in a mansion that, at the time, was the second largest private residence in the United States.

The importance of the LIBA to the affairs of the Biological Laboratory was apparent immediately. In 1926 LIBA members raised the funds needed to purchase 32.5 acres of land adjacent to the Laboratory. The following year, Mr. and Mrs. Acosta Nichols donated $12,000 for the construction of a laboratory in memory of their son George who had taken part in the Nature Study Class. These two activities—taking part in fund-raising drives and contributing to costs of construction—have remained key components of the LIBA's contributions to Cold Spring Harbor.

The enthusiasm of the LIBA members was demonstrated by the grand fund-raising gala the Marshall Fields held at Caumsett, their estate in Lloyd Harbor. It was a "Dutch Treat Dinner," in which some 100 friends of the Fields provided tables for more than 500 invited guests who paid $5 per ticket. The entertainment included dancing and china breaking, the latter being "… overseen by Mr. Vincent Astor."

LIBA operated the Biological Laboratory for 38 years until the reorganization in 1962 when the CIW closed the Department of Genetics.

(Left) Newspaper clipping and (middle) invitation to the grand fund-raising gala at Caumsett, Marshall Field's 2000-acre estate in Lloyd Harbor. (Right) Attendees included Fred Astaire (center).

The latter's assets were merged with those of the Biological Laboratory to create a new entity, the Cold Spring Harbor Laboratory. LIBA remained one of the institutions supporting the new entity. There was a change of name in 1991 when the Long Island Biological Association became the Cold Spring Harbor Laboratory Association to make clear the relationship between the two bodies. But the primary mission of CSHLA remains the same, to raise "… unrestricted annual support for research and education at the Laboratory." The Laboratory has been most fortunate to have had such supporters for more than 90 years.

A New Director of the Biological Laboratory

In 1924, the Biological Laboratory was in strong shape—financially and in terms of research and staff. It received a final payment of nearly $1800 from the Brooklyn Institute in March. Contributions from the Wawepex

Society totaled $1500, with an additional $1774 from other supporters. Tuition fees from 84 course participants and summer researchers brought in more than $4500, with nearly $2500 more from rentals of laboratory space and other facilities and equipment. Other receipts, including a small endowment income, brought the total income of the Laboratory to $21,144.22, and expenditures were held to precisely that amount.

The formation of LIBA was Davenport's swan song as head of the Biological Laboratory. With the Laboratory's fortunes secure, he could at last resign and focus on his greater passions, genetics and eugenics. Davenport remained director of the Carnegie Department, and when the transfer of the Laboratory to LIBA was complete, he turned over the reins to his son-in-law, Reginald Harris.

A suave, handsome, energetic young man, Reginald Harris had married Charles Davenport's daughter Jane in 1922. Harris was a graduate of Brown University, and in 1919 he had come to Cold Spring Harbor for the summer to work with C.C. Little. He then took a trip to South America, from where he plied Davenport with letters detailing his observations and thoughts on insects and the native South American peoples, even writing and sending Davenport a paper on eugenic possibilities in South America. By 1922 Harris was completing his dissertation research at Brown and wrote letters to his faculty advisors and other mentors, soliciting a job. One of these went to Davenport, who replied he might be able to find Harris a post at Cold Spring Harbor, "perhaps in eugenics."

Within a few months, however, by the spring of 1923, Davenport had recommended Harris not for a research position but for the directorship of the Biological Laboratory. In 1924, Harris received his Ph.D. from Brown University and was named acting director of the newly reorganized Biological Laboratory at Cold Spring Harbor. Upon the formal transfer of the Laboratory to the Long Island Biological Association in January of 1924, the Board of the Association confirmed Reginald Harris as the director (Davenport was kept on as Secretary). A new era was about to begin.

Reginald Harris.

Reginald Harris Carves Out a New Niche for the Laboratory

Harris immediately plunged into the job, for now the directorship was a full-time position, and Harris was the first man to head the Biological Laboratory who had no other appointment elsewhere. The first two directors, Bashford Dean and Herbert Conn, had had university appointments and came to Cold Spring Harbor only in the summer. Davenport was on the staff of the University of Chicago until the Station for Experimental Evolution opened in 1904. Its year-round program brought him to Cold Spring Harbor full time, but even so the Biological Laboratory had suffered as the SEE and the Eugenics Records Office absorbed more and more of his attention.

Reginald Harris at his microscope.

Harris needed to justify his permanent position as director of the Laboratory, and he did so by beginning the slow process of turning the Biological Laboratory into a year-round research laboratory, even if at first this was more image than substance. Under Harris, research would take precedence among the Laboratory's activities. "The foundation of biology is research," he wrote:

> Without scientific observation and inquiry there would be no zoology, or botany, no rational medicine, agriculture or animal husbandry. Continued advance in the knowledge of life, its continuity, diversity, development, differentiation and pathology, is directly dependent upon continued research.... Without research there can be no instruction in biology, for there can be no facts or sound theories to teach.

Harris's first step was to provide for expansion of the Biological Laboratory. In his first annual report (1924), Harris wrote that the purchase of new land was most urgent: "It may be said that upon its acquisition depends the future of the Laboratory." The land would be used for a "... much needed Research Laboratory... . The immediate demand is insistent." Where was the money to come from? From LIBA: "It is for the Association to say whether the Laboratory shall be relegated to an inferior position due to inadequate equipment or whether it shall be able to play an important role in the development of biology." Fortunately, LIBA was up to the task. Marshall Field, William Matheson, Mortimer Schiff, and W.K. Vanderbilt each contributed $10,000 to a fund that was used to purchase 32.5 acres of land adjacent to the Laboratory. By 1926, the new laboratory, built to the north of Hooper House and named for Davenport, was in use.

In 1927, Harris formed a scientific advisory committee, headed by Joseph Hall Bodine of the University of Pennsylvania. The committee recognized that the Biological Laboratory's reputation depended primarily on its research activities and urged the expansion of laboratory space.

W.K. Vanderbilt writes to Davenport pledging to match Mortimer Schiff's gift
of $10,000.

Harris began major renovations of the Wawepex Building in order to
convert it from a lecture hall and teaching lab into a research laboratory. Building continued in 1927 when construction began on another
laboratory, funded by a donation of $12,000 from Mr. and Mrs. Acosta

Davenport laboratory, 1926.

Nichols and to be named for their son George Lane Nichols. A third building, costing $12,000, was constructed in 1928, to house the biophysics program Davenport had advocated. The donor was Mrs. Walter B. James, and the building was named the James Laboratory, in honor of her late husband who for 26 years had been a member of the Board of

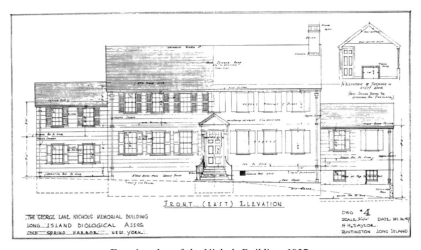

Exterior plan of the Nichols Building, 1927.

The Dr. Walter B. James laboratory for Biophysics, 1929.

Directors of the Biological Laboratory, and who, at the time of his death in 1927, was the President of the Long Island Biological Association. The James Laboratory was added to over the years but the original 1929 building can still be seen.

An unusual building was added to the Biological Laboratory when, in 1930, Harris bought the old (1906) Cold Spring Harbor Fire Department's firehouse for $50. It was floated across the Harbor on a barge and placed just north of the Davenport Laboratory. This was not its final resting place—in 1986 it was moved further north so that an addition could be made to the Davenport Laboratory. The firehouse was converted in to living quarters for staff and visitors.

The building program permitted what was by far the greatest change, even if it was largely symbolic, in the Laboratory during Harris's first years: the switch to year-round research. The use of the facilities throughout the year would be economical, Harris said, because it avoided the cost of maintaining facilities that were shut down for three-quarters of the year and because becoming a year-round institution would undoubtedly boost the Laboratory's image. In 1928, Harris boasted that since

The firehouse approaches the grounds of the Biological Laboratory.

1923 the annual number of researchers at the Laboratory had increased from three to 19. And although nearly all of these were summer visitors, it nevertheless indicated a dramatic increase in activity. Harris could also now encourage scientists from other institutions to spend sabbatical years at Cold Spring Harbor, instead of just the summer.

The "Hot" Field of Biophysics Comes to the Biological Laboratory

In 1928, Harris established a research program in biophysics and at last made the Laboratory a year-round research institution. The biophysicist, armed with electrodes, voltmeters, and conductivity meters, explored the chemical and electrical basis of excitable membranes and cellular activity. Harris's first foray into biophysics was the appointment of biophysicist W.J.V. Osterhout of the Rockefeller Institute to the LIBA board of directors in 1927 and as an honorary staff member in 1928. Osterhout urged Harris to expand the Laboratory's biophysics effort, and Harris complied by establishing a special Advisory Committee for General Physiology and Biophysics, which recommended Hugo Fricke as the Laboratory's first full-time scientist in this area.

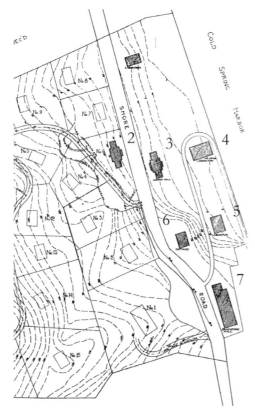

1926 map showing the buildings along Shore (now Bungtown) Road. (Current names: 1, Davenport; 2, Williams; 3, Hooper; 4, James; 5, Wawepex; 6, Osterhout; 7. Blackford.)

Fricke came to Cold Spring Harbor from the Cleveland Clinic Foundation, where he worked on the effects of X rays on cells and the properties of the cell membrane, research he continued at the Biological Laboratory. He moved into the new James Laboratory, built to the exacting needs of a biophysics laboratory. Its concrete structure and siting on the hillside away from the water helped ensure it would be dry and quiet, both electrically and acoustically. Fricke worked with Howard Curtis, who came to the Biological Laboratory in 1932. Together they published

Hugo Fricke.

several papers on the electrical impedance of cell membranes, at the time one of the few sources of data that could be used to try to deduce the structure of the cell membrane. In 1934, Fricke was joined by Eric Ponder, a distinguished investigator of the properties of the red blood cell. Ponder was an authority on hemolysis, an important clinical topic, and the author of the standard textbook *Hemolysis and Related Phenomena*.

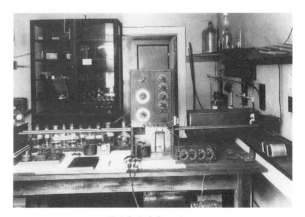

Fricke's laboratory.

Electric Impedance of Suspensions of Yeast Cells

THE electric impedance of suspensions of yeast cells, suspended in solutions of electrolytes, has been measured as a resistance, R, and a parallel capacitance, C, with a Wheatstone bridge. In Fig. 1 are shown C and R, as functions of frequency, for a 63 per cent suspension of yeast cells in a 0·1 per cent sodium chloride solution. The form of the curve for C is interesting, particularly when it is compared

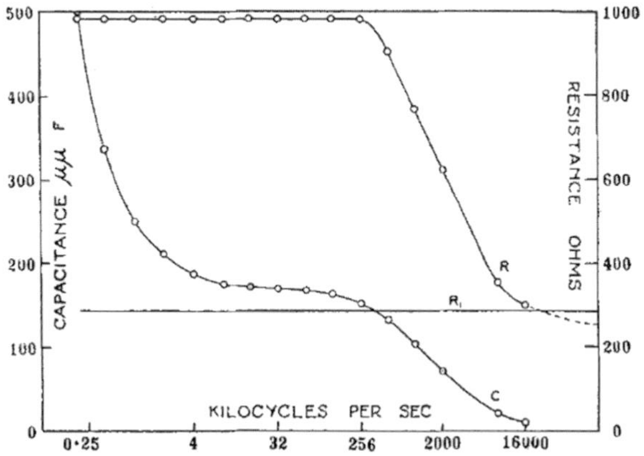

FIG. 1.   Resistance (R), capacitance (C) and resistance of suspending fluid (R_1) for a 63 per cent suspension of yeast in 1 per cent sodium chloride*.

HUGO FRICKE.
HOWARD J. CURTIS.
Dr. Walter B. James Laboratory for Biophysics,
Biological Laboratory,
Cold Spring Harbor, Long Island, N.Y.

A 1934 paper by Fricke and Howard Curtis, typical of Fricke's research at the Biological Laboratory. It appeared in Nature*, Vol. 134, page 102.*

A New Research Program in Physiology and Endocrinology

In the 1920s, quantitative biology also meant physiology, and Harris built a substantial research effort in physiology, with both existing and new staff. By 1929, he had assembled a second special scientific advisory committee, on reproductive physiology. Chaired by John W. Gowen

Alice and Wilbur Swingle.

DYING PATIENT RESTORED TO HEALTH WITHIN 2 DAYS

Injection of Hormone Brings Man New Lease of Life

Cold Spring Harbor, N. Y., Dec. 5.
—(AP)—Restoration of a man from death's door to apparently perfect health in forty-eight hours was revealed today at the Biological Laboratory.

The man saved, a patient at the Mayo Clinic at Rochester, Minn., had Addison's disease.

At the clinic a purified hormone was injected directly into the veins of a 39-year-old farmer by Dr. Leonard G. Rowntree and Dr. Carl H. Greene.

"Before its use," they reported, "the patient showed evidence of failing circulation. Within forty-eight hours, he had taken a new lease of life and he appeared to be in perfect health."

A 1930 newspaper headline greets the success of Swingle and Pfiffner's hormone in treating an Addison's disease patient.

of the Rockefeller Institute, it included John Hammond of Cambridge University (England), George W. Corner of the Rockefeller Institute, Alan Sterling Parkes of the University of London, and Harris himself.

"Internal secretions"—hormones—were an exciting area of physiological investigation and represented a rigorous, mechanistic way to examine living systems. Hormones had specific physiological targets, they could be manipulated experimentally, and their effects could be measured quantitatively. The summer investigations of Wilbur Willis Swingle of Yale University on the adrenal cortex and the nature and functions of its hormones were especially exciting.

Although it was known that the medulla, or internal portion, of the adrenal gland produced adrenaline, the function of the outer, cortical area remained mysterious. In 1927, Swingle, whose home institution was the University of Iowa, showed that the cortex was essential for the kidney to neutralize acid in the urine, and he began to purify extracts of adrenal cortex to identify its active substance. His goal was to develop drugs to treat victims of Addison's disease, believed to result from dysfunction of the adrenal cortex. In 1928, Swingle moved to Princeton. He did not come to Cold Spring Harbor that summer, but Harris sent staff member Joseph John Pfiffner, a chemist who had been working with Swingle, to Princeton to continue their collaboration. In 1929, Swingle returned to Cold Spring Harbor and succeeded in isolating the active principle of the adrenal cortex. He showed that the cortical extract was effective in keeping alive animals that had had their adrenal glands removed. In 1930, his extracts were tested on Addison's patients at the Mayo Clinic, immediately relieving the symptoms of the first patient on whom it was tried.

Other physiology studies reported in 1928 included the work of William Salant, who had been appointed to the staff but was on leave of absence that year. He investigated the pharmacology of ergotamine and the effect of the sex glands on general resistance to various drugs.

Reginald Harris himself continued his studies in the physiology of reproduction, in particular with respect to the consequences of complete removal of the ovaries in pregnant female mice.

Summer Visitors Continue to Play a Major Role in the Life of the Biological Laboratory

Despite Harris's emphasis on the Biological Laboratory as a year-round research institute, the change from the Davenport days was less dramatic than Harris painted it. Research was indeed now emphasized over teaching, but although Harris made much of the research program in physiology, nearly all research continued to be done by summer investigators. The Laboratory remained largely a summer institute, with a skeleton crew of staff and scientists during the winter. In 1930, for example, Harris listed 24 researchers as "Laboratory Staff" but only four—Harris, Fricke, Salant, and Wharton—were reported as being in residence throughout the year. However, the summer investigators were very useful. Many were or were becoming eminent, and Harris incorporated their work in the Biological Laboratory annual report, creating the impression of a vigorous research program that would, he hoped, attract more researchers and further build the Laboratory's reputation.

The biophysics program, for example, was much enhanced by summer researchers. These included H. Davson and J.F. Danielli, famed for their eponymous model of the cell membrane; J.Z. Young, known for his discovery of the squid giant axon; N. Rashevsky, for theoretical biology; and Dorothy Wrinch, for her work on protein structure, particularly her controversial "cyclol" theory.

Although research was now the most important aspect of the Laboratory, summer courses remained a part of Harris's vision. In 1932, five courses were offered: Field Zoology, General Physiology, Surgical Methods in Experimental Biology, Plant Sociology, and Bryology.

John Randall and James Danielli at the 1949 Symposium on Amino Acids and Proteins. *(Randall was head of the MRC Biophysics Unit at King's College, London, and later the boss of Rosalind Franklin and Maurice Wilkins.)*

Dorothy Wrinch on the Blackford balcony at the 1940 Symposium on **Permeability and the Nature of the Cell Membranes.**

The Biological Laboratory's Finances Improve

Visitors brought not only intellectual stimulation to the Laboratory but economic stimulation as well. Visiting researchers were charged a tuition that included laboratory privileges for the summer, one summer course, and lectures. Harris raised the tuition from $50 to $60 in 1925, to $70 in 1926, and $75 in 1928. That enrollment continued to climb during these rises suggests the courses were found valuable. Those attending for research only paid a $50 "table fee," and institutions could purchase a table for an entire year for $100 per person. This system was based on the "table system" developed at the Naples Zoological Station in the 19th century. Meals cost $7.75 per week, dormitory rooms ran $2.75 to $4 per week, and tent space was available for $2.75 per week. Tuition in 1928 brought in more than $4100 and was second only to dues and contributions ($16,875) as a source of income.

With the increasing subscriptions, a large and vigorous board of directors, and more members and contributors than ever, the Biological Laboratory was not only in charge of its own destiny but was richer than it had ever been. The administrative and scientific improvements complemented each other: As the Laboratory gained in prominence, it became easier to raise money and attract new members; conversely, the more money Harris had to build, maintain, and improve laboratory facilities and accommodations, the easier it was to attract top-quality students and scientists.

As the scientific enterprise grew more expensive, Harris searched for other sources of money. He appealed to the Carnegie Corporation and the Rockefeller Foundation, and, in 1930, both institutions came through with sizable grants. Harris was pleased, not only for the financial support, but also for the demonstration of confidence. The Rockefeller Foundation gave $20,000 in unrestricted research support, and the Carnegie Corporation gave $15,000 to support the adrenal cortex work of Swingle and Pfiffner. The National Research Council, one of the few

Patented Mar. 23, 1937   2,074,492

UNITED STATES PATENT OFFICE

2,074,492

MEDICINAL COMPOUND

Wilbur Willis Swingle and Joseph John Pfiffner,
Princeton, N. J., assignors to Parke, Davis &
Company, Detroit, Mich., a corporation of
Michigan

No Drawing. Application August 28, 1930
Serial No. 478,558

7 Claims. (Cl. 167—77)

The invention relates to an extract obtainable from mammalian suprarenal glands and a method of preparing the same. More particularly the invention pertains to an extract from the suprarenal cortex containing in concentrated form the active principle or hormone of the adrenal cortex. chloride, etc. Under proper mechanical conditions the comminuted gland tissue can be extracted directly with a water-insoluble solvent like benzol but on a laboratory scale we prefer the sequence of two extractive agents, one being an organic solvent soluble in water and the sec-

The Swingle patent.

federal sources for research money, gave money to help buy expensive equipment for Hugo Fricke's biophysics work. The Laboratory even benefitted from "technology transfer" when the pharmaceutical firm of Parke, Davis bought Swingle's adrenal cortical extract for treating Addison's disease. Swingle gave the royalties, amounting to $10,000, to the Laboratory to support further research. So, even as the national economy was sliding into depression during the first years of the 1930s, the Laboratory remained reasonably secure.

By 1933, however, the Depression had permeated even the financially sheltered coves of Long Island's North Shore: Member contributions to the Laboratory sank to their lowest level since 1926. Funds continued to be tight at the Laboratory as elsewhere throughout the Depression. Harris provided a thoughtful if melodramatic analysis of the situation. In hard times such as these, he wrote, a man, or an institution, has three options:

One is to commit suicide. Another is to change one's mode of life and one's aims, temporarily and permanently, in order that relative security and peace of mind may be achieved, though, perforce, on a much more modest basis than had originally been planned. The third is to build with broader foundations for the future, since, knowing the exigencies and upsets of life, one is only the more determined to make one's life of the greatest possible satisfaction and value to himself and to his fellows.

All three of the possibilities have been, and are being, adopted in these times by men, and by organizations and institutions of men.

The Biological Laboratory is no exception. A choice must be made.

Harris's choice, not surprisingly, was to pursue his vision of quantitative biology, broaden his foundations, and plan for the future. Private research laboratories, he argued, had a special role to play in such times. Not only did he assert that research could be performed at a research laboratory for about half what it would cost at a university, but also that a higher grade of work would result, because research laboratory scientists would not be diverted from their experiments by teaching and administrative duties. The best scientists were not needed at the university, he argued. Teaching and research skills are by no means perfectly correlated and teaching responsibilities detract from research efforts. The best return for the research dollar, he concluded, was to be had at a private institution such as Cold Spring Harbor, where scientists could focus 100% of their time on experiments.

Harris also recognized that collaboration and interaction among scientists at universities and research laboratories were essential to the advance of scientific knowledge. In this period of nationwide financial hardship, Harris devised an ambitious plan that would promote his vision of quantitative biology and elevate the position of the Biological Laboratory intellectually and financially. In 1934, the first Cold Spring Harbor Laboratory Symposium on Quantitative Biology was held.

WILBUR WILLS SWINGLE AND JOHN PFIFFNER ISOLATE CORTIN, WHICH IS QUICKLY USED TO TREAT ADDISON'S DISEASE

Summary

Swingle opened the door to the adrenal corticosteroids with his isolation of cortin, which was immediately shown to be an efficacious treatment for Addison's disease. Cortin more properly is an adrenal cortex extract containing a mixture of hormones—the corticosteroids. Further research by other investigators on the "Swingle–Pfiffner extract" separated the various components of this extract—glucocorticoids, including cortisol, and mineralocorticoids, including aldosterone—and produced treatments for many conditions, most notably for arthritis.

The Discovery/The Research

In 1849, Thomas Addison described the association between degeneration of the adrenal glands and the disease now named after him. Addison's disease is primary hypocortisolism, or adrenal insufficiency, and its deficient production produces chronic elevation of ACTH resulting in a variety of symptoms.

By the early 20th century, "internal secretions," or hormones were starting to be isolated. Quantitative biology in the 1920s meant physiology, and this included the search for hormones. To this end, in 1924, Reginald Harris, appointed Swingle to the summer research staff.

In 1937, Swingle reported he was trying to isolate the active factor in adrenal cortex extracts. The next year, Swingle accepted a full-time position at Princeton University and conducted research at both institutions. To aid Swingle in separating out the adrenocortical factors, Joseph John Pfiffner, a chemist at Parke, Davis & Company, was assigned to work with Swingle in 1929. They soon managed to produce an extract depleted of adrenaline; treatment of suprarenalectomized cats with their extract were "restored to normal" and daily injections kept them healthy.

In 1930, they eliminated adrenaline from the extract and they described a "cortical hormone" in the 1930 paper "The treatment of patients with Addison's disease with the 'cortical hormone' of Swingle and Pfiffner." This paper also described the treatment of an Addison's disease patient at the Mayo Clinic, in Rochester, Minnesota. Leonard G. Rountree and C.H. Greene administered the extract to four patients and reported that their symptoms improved after 2–3 days. Swingle and Pfiffner fractionated the extract and were able to remove most of the epinephrine, which had been causing severe irritation at the injection site.

Wilbur Willis Swingle in 1937.

Continued

The patent included Swingle's and Pfiffner's names. A 1931 *New York Times* article detailed a talk by Frank A. Hartman at the American Chemical Society meeting in which he described the "Swingle–Pfiffner extract." He reported that the "cortin extract at present is one of the world's most precious substances. It required the adrenal organs of 180 cattle to yield an amount the size of a pinhead, enough for just one day's administration."

1933 was Swingle's last year of research at Cold Spring Harbor. And with the death of Harris in 1934, the endocrinology program at Cold Spring Harbor ended.

Significance

Addison's disease is fairly rare—approximately 120 cases per million people in Western countries. The most famous case of Addison's disease is that of President John F. Kennedy.

Later research showed that cortin contained many steroids, termed adrenal corticosteroids. Edward Kendall of the Mayo Clinic isolated cortisol, and then in 1936 he prepared corticosterone in crystalline form. In 1937, a synthetic form of corticosterone was produced by Tadeus Reichstein in Basel, and it was used for treatment beginning in 1938. The 1950 Nobel Prize in Physiology or Medicine was shared by Reichstein, Kendall, and Philip S. Hench for their discovery of adrenal corticosteroids. Kendall graciously acknowledged Swindle and Pffifner in his Nobel lecture.

The Scientist

Swingle was born January 11, 1891, in Warrensburg, Missouri. He earned his Ph.D. at Princeton University in 1920 with the thesis topic "The germ cell cycle of Anurans: 1. The male sexual cycle of *Rana catesfrava*." From 1920 to 1926, he was Professor of Zoology at Yale University. While on the summer research staff at Cold Spring Harbor, Swingle was also Professor of Zoology at the State University of Iowa (1926–1929). After his retirement from Princeton in the late 1960s, he did research at the New Jersey Neuro-Psychiatric Institute. During his research career, he published more than 200 research papers. He died May 20, 1975, in Long Beach, California.

—*J.C.*

7

The Symposia on Quantitative Biology

Origins

Harris's ambitions for promoting quantitative research in biology extended far beyond the Cold Spring Harbor campus. As early as 1931, Harris had written that the Laboratory "… continues to act as a clearing house for scientific methods and ideas … ." This function was put on a more formal footing when, in 1933, he initiated

> … a plan whereby each summer a small group of investigators will be brought together covering some important phase of biology… . The general problem receiving special attention was be changed from year to year, being chiefly concerned with quantitative biology, particularly those fields related to the exact sciences, mathematics, physics and chemistry, and intermediate divisions.

Harris elaborated on this plan in the 1934 annual report:

> Now it happens in modern biological research that the problem of a given biologist, or group of biologists, may have much more in common with that of a given chemist or physicist, or a small group of either, than with that of a second group of biologists… . Yet the meetings of the various learned societies in this country still fail to take this into account… . The primary motive of the conference symposia is to consider a given biological problem from its chemical, physical and mathematical, as well as from its biological aspects.

Harris's motives are unlikely to have been entirely altruistic. The "conference-symposia" were a part of Harris's larger plan to professionalize

Reginald and Jane Harris.

the Laboratory, to make it a world leader in biological research and dissemination of data, and to make his own mark on biology. Besides bringing the Biological Laboratory prestige through the event itself and circulation of its proceedings, it would encourage scientists to spend the summer at Cold Spring Harbor doing experiments.

The First Symposium

Not surprisingly, given Harris's interest in biophysics, the first Symposium, held in the summer of 1933, was on *Surface Phenomena*. There were papers, among others, on the electrical properties of membranes and nerve cells, on using measures of osmotic pressure to investigate the physical chemistry of cell membranes, and on oxidation–reduction reactions. There were only 29 scientists in attendance, all drawn from North America—28 participants from the United States and one from

> COLD SPRING HARBOR SYMPOSIA
> ON QUANTITATIVE BIOLOGY
>
> ୯୨
>
> #### Introduction
>
> At the first meeting, held July 1, 1933, Dr. Harris made the following remarks:
>
> "In opening this conference it seems desirable to state briefly why the Biological Laboratory at Cold Spring Harbor has invited you to join together in experimental work and in conference this summer.
>
> "The officers of the Laboratory are interested in the development of an institute in which biologists, chemists, physicists and mathematicians will cooperate in the further opening, and beneficial use, of the vast territory of quantitative biology.
>
> "The initial step toward the accomplishment of this aim was taken, in respect to the all-year work of the Laboratory, in 1928, when Dr. Hugo Fricke, a physicist, was appointed to establish and direct our laboratory for biophysics. The second move has brought into being the conference which now begins.
>
> "The present meeting is the inauguration of a plan whereby each summer a group of mathematicians, physicists, chemists and biologists, actively interested in a specific aspect of quantitative biology, or in methods and theories applicable to it, will be invited to carry on their work, to give lectures and to take part in symposia at the Laboratory. A given group in residence here will necessarily be relatively small, but members of the group will be chosen with the aim that every important aspect of a particular subject is adequately represented from the physical and chemical, as well as from the biological point of view; and that the whole span of a subject, from theories of physics to application to medicine, is covered.

Part of Reginald Harris's introduction to the first Symposium.

CONTENTS

Contents of the first Symposium volume.

Canada. However, two foreign scientists who did not attend contributed papers to the Symposium volume—A.V. Hill (Nobel laureate, 1922) contributed a paper on nerve conduction, and T. Svedberg (Nobel laureate, 1926) sent in a paper on ultracentrifugation. Contributions of this kind by prestigious scientists helped elevate the status of the Symposia. One-half of the papers appeared within 2 weeks (!) as "preliminary" communications in *The Collecting Net*, the publication of Woods Hole.

The Brooklyn Eagle *reports on the 1933 Symposium.*

These early Symposia were not the frenetic, action-packed meetings that we know today. Instead a small group of scientists stayed for 1 month at the Laboratory carrying out research, and additional participants came periodically to present papers and to participate in the Symposium. For example, the 1936 Symposium began on Tuesday, June 23 and ended on Friday, July 24. Of the 49 scientists listed as participants

Participants in the 1936 Symposium relax by the harbor.

in the Symposium volume, 30 were "… in residence at Cold Spring Harbor during all, or an appreciable part, of the five weeks' period." During this period, these full-time participants carried out experiments at the Laboratory. There were 35 presentations listed in the program, and the maximum number of presentations on any day was three and usually only one. The day's program began at a civilized time—10:20 a.m. (The length of the Symposia was shortened to 2 weeks in 1941; to 8 days in 1948; and to 5 days in 1997.)

Harris had written that he expected that the meetings "… quite likely may be very useful to biological research." That the first Symposium had filled an important function was evident when the Rockefeller Foundation provided $5000 toward the cost of the second symposium. The Foundation continued to support the Symposia in the years leading up to World War II, providing the remarkable sum of $10,000 in both 1938 and 1939. Following the war, the Rockefeller Foundation helped the Symposia recover by providing grants of $12,000 in each of 1947 and 1948, and a further 5-year grant of $25,000 for the period 1958–1962.

Participants in the 1950 Symposium crowded into the Blackford Assembly Room (now the Racker Room).

The 1978 Symposium in Bush Auditorium.

Other invaluable support came from the Carnegie Corporation. which provided funding for 13 consecutive years, from 1948 through 1960. The Corporation also made a most important donation in 1950 when it contributed $100,000 to the cost of building a new lecture hall. The Bush Lecture Hall, named for Vannevar Bush, President of the Carnegie Corporation, provided much needed space (with air conditioning), symposium participants having long overflowed from Blackford's Assembly Room. Inexorably the numbers attending the Symposia grew, and by the early 1980s Bush Lecture Hall was inadequate. A new state-of-the-art auditorium, named for Oliver and Lorraine Grace, was built and used for the first time for the historic 1986 Symposium on *The Molecular Biology of* Homo sapiens.

The "Dusky Red Books"

The proceedings of the Symposium were published each year, bound in distinctive dark red covers. They were for many years a primary source

Oliver and Lorraine Grace Auditorium in 2003.

of information, and Sydney Brenner recalled how, when a medical student in 1943, "they were the main books which sustained me" in his isolation in South Africa.

Harris's goal was to publish the volume "as soon as possible" after a meeting, and that meant in the same calendar year. This was a heroic undertaking in the days before e-mail and word processors. Each volume contained papers presented during the meeting and, in the early days, contributions from scientists who had not participated in the Symposium. An outstanding feature of the Symposia volumes was the inclusion of the discussions that followed each paper. These were substantial additions to the submitted papers—Harris added 30,000 words of discussions to the first volume and more than 60,000 words to the second volume. Unfortunately, as the number of contributions grew, the discussions were dropped.

The volumes cost $4.50 in 1939 and soon provided a significant income for the Biological Laboratory; in 1941, for example, sales of the Symposia volumes accounted for 10% of the Laboratory's income. Sales of the Symposia volumes were even more important in the early 1960s when the

Laboratory was in severe financial problems. As John Cairns recounts, the 1963 Symposium volume, *Synthesis and Structure of Macromolecules,* became a "runaway bestseller." The advance sales "paid off most of the Laboratory's more pressing debts and enabled us to survive even when, in many people's minds, we had been given up for lost."

A fascinating feature of the Symposia volumes was introduced by Demerec in 1949, when he included a page of informal snapshots of participants in the *Amino Acids and Proteins* Symposium. These photographs provide an unrivaled gallery of many of the most influential scientists in the topic of each Symposium.

Kitty Brehme Warren and J.C. Foothills

Kitty Brehme, a *Drosophila* geneticist, was at Hofstra College, just a few miles from Cold Spring Harbor. She had prepared Calvin Bridges's *The Mutants* of Drosophila melanogaster for publication after Bridges's untimely death at the age of 49 years. Brehme had worked with Milislav Demerec, and he enlisted her to edit the 1955 and 1956 Symposia volumes.

J.C. Foothills of the Tennessee Intermountain College, Nazareth, Tennessee was listed as a participant in both Symposia. Foothills must have had wide-ranging tastes in science as the two Symposia were on rather different topics. The 1955 Symposium was on *Population Genetics* and the 1956 Symposium was on *Genetic Mechanisms: Structure and Function.* The index of the 1955 volume leads to a page in Motoo Kimura's paper full of the most abstruse equations, but inexplicably Foothills's name does not appear on the page. Foothills is also cited in 1956, but this time the page is blank. And although Foothills is listed in the caption to a photograph of Kimura, Emanuel Hackel, and Ernst Mayr, he or she must have bent down to pick something up just at the moment the photograph was taken. The entry in the index gives the game away to those in the know. It reads "FOOTHILLS, J.C., in" followed by the page number.

SOME SYMPOSIUM PARTICIPANTS

Top row: A. Keston, S. Udenfriend, F. Sanger; D. C. Hodgkin, A. L. Patterson; R. Benesch, H. J. Teas, A. H. Doermann.
Second row: D. M. Wrinch, B. W. Low; E. Brand, J. G. Kirkwood; B. P. Kaufmann, P. A. Abelson.
Third row: M. Levy; A. Marshak, A. Rothen, I. Fankuchen; S. Emerson, R. Herriot.
Bottom row: M. L. Anson, S. Simmonds, J. A. V. Butler; M. Demerec; C. Fromageot, V. Menkin, J. S. Fruton.

The page of snapshots from the 1949 Symposium Amino Acids and Proteins.

Brehme was well known for her mild swearing, particularly the colorful "Jesus Christ in the foothills." So J.C. Foothills assumed a life of his own, a literary example of an "Easter egg," and a permanent place in the most prestigious series of publications in molecular biology and genetics.

Biophysics Continues to Be the Topic

Harris's Scientific Advisory Committee noted "... with satisfaction the way in which the first conference-symposia on quantitative biology developed..." and recommended that they should continue. One feels that Harris needed no encouragement to undertake a second meeting, *Some Aspects of Growth,* in July of 1934. For Harris, growth was just the sort of problem he wanted to tackle. As he wrote in the introduction to the Symposium volume, "it is a very complex phenomenon ... the more complex the problem, the more the biologist must use mathematics, physics and chemistry, and the more valuable cooperation with representatives of these several sciences becomes." The number of participants had almost doubled to 58 and included Harold Urey, who was awarded the Nobel Prize in Chemistry later in the year, and Bill Astbury who a few years later, with Florence Bell, took the first X-ray photographs of DNA.

Harris demonstrated that the Symposium had already achieved his goal of examining a "... given biological problem from its chemical, physical and mathematical, as well as from its biological aspects" when he listed the disciplines of the participants:

mathematicians	2	physicists	5	biophysicists	5	physical chemists	2
chemists	2	biochemists	1	physiological chemists	1	physiologists	8
pharmacologists	2	biologists	8	zoologists	7	anatomists	3
cytologists	1	embryologists	1	geneticists	7	bacteriologists	2

He pointed out that at a large meeting there would be little opportunity for these participants to attend each other's talks as they would be distributed among at least nine disciplinary sections. Harris also noted, although he did not want to stress the fact unduly, that one of the chemists received the Nobel Prize that year, an indication, perhaps, of the "… type of men taking part in the conference-symposia." And the participants were indeed predominantly men, a reflection of the time. There were, however, at least three women participants in the 1934 Symposium.

The third Symposium on *Photochemical Reactions* exemplified the biophysical approach to biology, and the meeting began with a series of papers dealing with physical aspects of the interactions of light with molecules. One paper was by Dean Burk and Hans Lineweaver on "The kinetic mechanism of photosynthesis," making use of their publication the previous year of one of the most highly cited papers in biology, on enzyme kinetics.

This was the last Symposium that Harris organized. It was a tragedy for his family and for the Biological Laboratory when Harris died of pneumonia on January 7, 1936, at the young age of 38 years. It was said that the strain of editing the Symposia volumes contributed to his untimely death. He was succeeded by Eric Ponder, who had come to the Biological Laboratory in 1934. As Ponder was a biophysicist, it is not surprising that the five Symposia overseen by Ponder continued the theme of biophysics set by Harris. Ponder also included a meeting on *Internal Secretions*, reflecting the work being done in the Biological Laboratory on prolactin and adrenocorticotropic hormone.

Genetics Becomes the Dominating Theme

The choice of Symposia topics is made by the directors of the Laboratory and, not surprisingly, have generally followed the interests of the directors. So, an abrupt shift in symposium topics came in 1941 when Milislav

Demerec became director of both the Carnegie Institution's Department of Genetics and the Biological Laboratory. It was during Demerec's tenure as director that the Laboratory established its reputation as a center for research in microbial genetics.

The first of Demerec's symposia was the 1941 landmark meeting on *Genes and Chromosomes: Structure and Organization,* and the first of the post–World War II meetings (1946) was on *Heredity and Variation in Microorganisms.* These were followed by related meetings in 1951, 1953, 1956, and 1958. These symposia heralded the development through the 1950s and 1960s of what became known as "molecular genetics" and the research presented at the Symposia helped define the field.

Demerc's successors—Edwin Umbarger, Arthur Chovnick, and John Cairns—all contributed to this theme. Umbarger organized the 1961 Symposium, where Jacob and Monod described their operon model, and Chovnick organized the 1963 Symposium on *The Synthesis and Structure of Macromolecules.* By 1966, John Cairns was director and he organized one of the most famous of the Symposia, on

(Left to right) *Ed Lewis, Carl Lindegren, Al Hershey, and Joshua Lederberg at the 1951 Symposium on* Genes and Mutations.

The Genetic Code. It was, as Francis Crick wrote "... an historic occasion" and a fitting culmination to the series of Symposia that had begun 20 years earlier.

(Left to right) *Gobind Khorana, Francis Crick, and Marianne Grunberg-Manago at the 1966 Symposium on* The Genetic Code.

The series of classic Symposia on molecular genetics

1941	Genes and Chromosomes: Structure and Organization
1946	Heredity and Variation in Microorganisms
1947	Nucleic Acids and Nucleoproteins
1949	Amino Acids and Proteins
1951	Genes and Mutations
1953	Viruses
1956	Genetic Mechanisms: Structure and Function
1958	Exchange of Genetic Material: Mechanism and Consequences
1961	Cellular Regulatory Mechanisms
1962	Basic Mechanisms in Animal Virus Biology
1963	Synthesis and Structure of Macromolecules
1966	The Genetic Code

The Scope of the Symposia Widens

Even during the heyday of Symposia on molecular genetics, many other topics were covered, and this became even more true from the late 1960s. Much as Harris had seen biophysics as the key to understanding biological processes, 30 years later the tools and techniques of molecular biology came to the fore. As a consequence, the Symposia topics became more diverse, drawing on neuroscience, immunology, cancer, cell biology, and genetics. Some themes are followed over years. For example, the 1936 Symposium on *Excitation Phenomena* was followed by seven other meetings on neuroscience some 11 years apart, and there have been five immunology Symposia, again at intervals of 11 years.

The list of notable speakers and findings presented at the Symposia is long. Taking some examples from the early years of molecular genetics:

Niels Jerne at the 1967 Symposium on Antibodies.

(Left to right) *Stephen Kuffler, Alan Hodgkin, and Sydney Brenner at the 1975 Symposium on* The Synapse.

Joshua Lederberg and Edward Tatum presented their data on bacterial recombination in 1947; Fred Sanger described his first results determining the amino acid sequence of insulin in 1949; in 1951, Barbara McClintock described her work on her talk on the *Ac-D*s system in maize; in 1953, Jim Watson made the first presentation of the DNA double helix at an open meeting; in 1958, Matt Meselson and Franklin Stahl described their eponymous experiment; in 1961, François Jacob and Jacques Monod presented their operon model; and, in 1966, Francis Crick presented a synthesis of what was known of the genetic code. Ten years later, the world of molecular genetics was turned upside-down with the first reports of RNA splicing by Rich Roberts and colleagues and by Phil Sharp and Susan Berget.

Lectures and Dinner Parties

In addition to the regular sessions, there are now two special lectures at each Symposium. The Dorcas Cummings Lecture began in 1978. Dorcas Cummings was a member of the Long Island Biological Association (LIBA) for more than 20 years, and on her untimely death in 1976, a

Francis Crick talks with LIBA members after his Dorcas Cummings Lecture at the 1990 Symposium, **The Brain.**

fund was established to support a lecture in her memory. The audience for the Dorcas Cummings lecturer are members of the Association and other neighbors of the Laboratory, together with scientists attending the Symposium. A more recent addition is the Reginald G. Harris Lecture, which was first held in 2000. It is given by a participant in the Symposium to the other participants and provides an opportunity for the lecturer to give a longer, more reflective treatment of a topic.

A unique contribution to the social events of the Symposium is made by the Association members. In the 1930s, during the early days of the Symposia, housing for participants was in short supply, and LIBA members, including the Tiffany and de Forest families, opened their homes to visiting scientists. This led to the tradition of LIBA members hosting dinner parties in their homes on Sunday night during the Symposia. In the 1950s, the tradition was ratified when LIBA's women's committee assumed responsibility for planning and coordinating the parties. In recent years, LIBA's successor, the CSHL Association, has taken on responsibility for arranging dinners for visiting scientists at the homes of neighbors and friends of the Laboratory. The dinners are now held following the

LIBA members carrying signs with their names collect their scientist dinner guests after Günter Blobel's Dorcas Cummings Lecture in 1995.

Dorcas Cummings lecture and provide an opportunity for the scientists to experience the hospitality of the Laboratory's supporters, and for the latter to interact with and ask pointed questions of scientists.

Reginald Harris left a legacy that has stimulated the generations of biologists that have come to Cold Spring Harbor each year to talk, argue, and gossip at the Symposium. He set the tone in his introductory remarks to the first Symposium in 1933:

> May I ask you, then, in your lectures, to give special consideration to theoretical and controversial aspects, that the discussion may be both significant and creative, and that the conferences may be of the greatest possible value not only to those of us who take part in them, but also to those who will have occasion to refer to them.

The internet is bringing changes to the Symposium—tweeting is now a fact of life—and we can be sure that Harris would have embraced every opportunity to make the Symposia even more "...centers of growth and dissemination of new methods and ideas in biology."

8

Cold Spring Harbor, 1930s–1940s

The Department of Genetics

Although the Biological Laboratory had undergone significant changes in the 1920s—the appointments of Harris as the new director and a new governing body, a shift to biophysics, and the introduction of the Symposium—the Department of Genetics had remained relatively stable. The major administrative change came in 1934 when Blakeslee became director on Davenport's retirement. However, the research programs initiated by Davenport continued largely unchanged during Blakeslee's tenure, and under his leadership the Department of Genetics established itself as a major center for genetics research.

Demerec Makes Cold Spring Harbor a Center for Drosophila Genetics

One part of Davenport's program that increased in importance was research on *Drosophila*. Charles Metz had come to the SEE as a *Drosophila* geneticist, but in the early 1920s he became fascinated by the cytogenetics of *Sciara coprophila,* an obscure fungus gnat, which he had found in one of Oscar Riddle's pigeon houses. By 1925, Metz had abandoned research on *Drosophila,* and in 1930, he left the CIW's Department of Genetics at Cold Spring Harbor for the CIW's Department of Embryology in Baltimore.

Metz's move from *Drosophila* to *Sciara* did not mean the end of *Drosophila* genetics in the Department of Genetics. He had already introduced a young colleague, Milislav Demerec, to the advantages of the

Blakeslee relaxes on Blackford balcony during the 1942 Symposium.

A young Milislav Demerec at his microscope with culture bottles containing fruit flies on the bench in front of him.

fruit fly for genetics. Demerec had come to Cold Spring Harbor in 1924, from Cornell, where he had completed his Ph.D. under the great maize geneticist Rollins Emerson. Demerec continued to work on maize but soon switched to studying variation in *Delphinium.* He found six genes that appeared unstable, but genetic analysis in the slow-breeding delphiniums with only two or three generations each year was going to be slow.

Metz, at that time still working on *Drosophila,* must have persuaded Demerec that the latter's investigations of mutability could be better studied in the fast-breeding fruit fly. Indeed, in *Drosophila virilis,* Demerec found just what he was looking for to continue his studies of mutable genes: a sex-linked recessive gene that was unstable and threw off many mutations in the course of an individual fly's development. The mutant was named *reddish*, since it changed the usual dark color of *Drosophila virilis* into a reddish yellow color. Demerec tried to map the position of the *reddish* gene on a *Drosophila* chromosome, but the data were inconsistent and inexplicable—he could not find its location.

Later, Demerec found two other mutants, *yellow* and *miniature*, that behaved like *reddish*. Demerec wondered if the instability of the *miniature* gene was caused by a chemical, by some aspect of the gene's structure (in the 1920s an unanswerable question), or by some other mechanism. Regretfully, Demerec had to admit that "It would be just a guess to try to answer these questions at present." It was not until 1965 that Melvin Green found unstable mutants in *D. melanogaster* and proposed that this phenomenon was similar to Barbara McClintock's controlling elements in maize. And it was not until 1982 that Gerry Rubin's molecular analysis showed that Green's conjecture was correct, and thus the anomalous behavior of Demerec's "unstable genes" was solved—56 years later!

Calvin Bridges at Cold Spring Harbor

By the end of the 1920s, studies in *Drosophila* dominated the world of genetics, and Demerec had established Cold Spring Harbor as the leading East Coast center for *Drosophila* genetics, with T.H. Morgan's group at the California Institute of Technology and H.J. Muller's at the University of Texas the centers in the West Coast and Southwest, respectively. Demerec was an organizer, and he set about using this skill to consolidate Cold Spring Harbor's position.

As a former maize geneticist, Demerec was familiar with the *Maize Genetics Cooperation News Letter*, first produced in 1929, which the maize community used to exchange information about strains, crosses, and techniques. Demerec thought that something similar was needed for *Drosophila* geneticists. In the fall of 1933, Demerec sent Theodosius Dobzhansky in Morgan's group a draft proposal to create a *Drosophila Information Service* (*DIS*) to communicate tips of the trade to the fast-growing and increasingly far-flung legions of young *Drosophila* geneticists.

Dobzhansky was enthusiastic but thought that Calvin Bridges, the technical master of the Morgan group, was the one who ought to participate. The problem was that although Bridges was brilliant, he was slow and unreliable. Dobzhansky thought Demerec should lead the effort and enlist Bridges' collaboration. Dobzhansky offered to act as a "tickler" to Bridges, to keep him moving and help him meet deadlines.

Demerec sent his proposal to Bridges, and Bridges was typically enthusiastic. He also proved to be typically unreliable. The project nearly foundered when Bridges failed to meet Demerec's deadlines. Dobzhansky swore he would "annoy Bridges continually, until he sends me to hell or further." One way to try to ensure Bridges stuck to the task at hand was to bring him to the Department of Genetics where he would be under Demerec's watchful eye.

In early 1934, Demerec invited Bridges out to Cold Spring Harbor for the summer, so that they could work on the *DIS* and Bridges would have a quiet place to pursue his work mapping the bands of the *Drosophila* salivary chromosomes. It was in 1933 that Theophilus Painter at the University of Texas (as well as Emil Heitz and Hans Bauer in Germany) had discovered the giant chromosomes found in cells of the *Drosophila* salivary gland. It was now possible to correlate the genetic linkage map of a chromosome with its physical structure. Bridges was a master of this technique, and his maps of the *Drosophila* chromosomes have never been bettered.

Bridges, freewheeling, terrifically good looking, an advocate of free love, with a passion for cars, was the polar opposite of the stolid, conservative, reliable Demerec. Despite their differences in temperament, an enduring friendship bloomed between Demerec and Bridges. This did not mean that Demerec approved of Bridges's lifestyle. If Bridges came, Demerec said, he must rein in his libertine tendencies:

Calvin Bridges at Cold Spring Harbor in 1935.

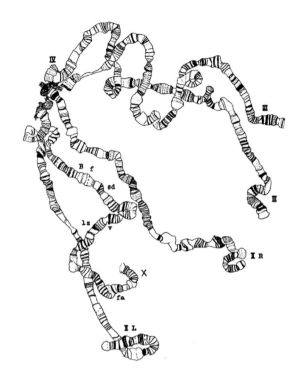

A figure from T.S. Painter's 1934 paper "A new method for the study of chromosome aberrations and the plotting of chromosome maps in Drosophila melanogaster," *showing an aceto-carmine-stained preparation of chromosomes in a salivary gland cell.*

As you know the Cold Spring Harbor biological community is a small closed group. Whatever is happening around is known to everybody. To avoid embarrassments, therefore, is it essential, in case you can come to make a gentleman's agreement that any mix-ups will be avoided with the female members of this community.

Bridges loved the idea and was willing to abide by Demerec's terms. He planned to motor east in a car of his own design and construction—"a hobby that's much safer than women." When the time came, the car was not ready, because, Bridges said, of delays in shipping of Japanese parts, so Bridges was forced to take more conventional transportation.

Blakeslee and Demerec arranged the space and facilities for Bridges to complete his masterful cartography. A suitable laboratory room was allocated for Bridges's sole use, on the upper floor of the Animal House. Windows were darkened, so that no extraneous light would produce glare and interfere with the visibility of even the weakest bands on the chromosomes. With superhuman patience, Bridges set himself to the exacting, tedious task of mapping, band by band, all four chromosomes of *D. melanogaster*. During the later years he had the assistance of his son Philip, then a student at Columbia University.

Hour after hour, day after day, summer after summer, in 1934, 1935, 1936, and 1937, Bridges slaved over this exacting chore. It was the most assiduous devotion to an arduous scientific task that many Cold Spring Harbor people had ever observed. Arthur Steinberg, who was working in the Animal House during one or more of those summers, recalled that periodically on the hour, Bridges would straighten up from his cramped hunch over the microscope, shrug the kinks out of his shoulders, raise

Lightning Bug Car Is Crash Proof

DR. CALVIN BLACKMAN BRIDGES, the Carnegie Institution of Washington geneticist, knows a lot about bugs, for he breeds them and studies their mechanism under the microscope. But his hobby is Safety First and car building. His three-wheeled "Lightning Bug" is said to be crash and carbon-monoxide proof. It is small and squatty yet perfectly streamlined, with tail-light and license plates recessed and the Pyralin windows flush with the body. The doctor's aim is to demonstrate what can be done to attain small car safety and roadability.

When tiring of breeding insects, Geneticist Bridges follows his hobby of Safety First. Out of his garage-laboratory came this 3-wheeled, streamlined car, the "Lightning Bug."

Bridges shows off his "Lightning Bug" to readers of Modern Mechanix *(1936).*

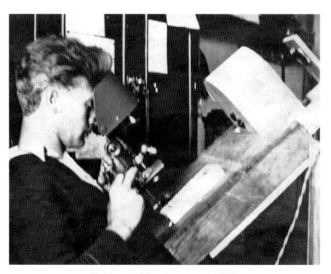

Bridges at his microscope, 1927.

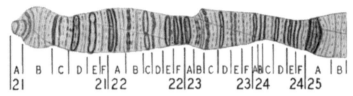

The left-hand tip of the second chromosome of Drosophila, *from Calvin Bridges's 1935 paper "Salivary Chromosome Maps: With a Key to the Banding of the Chromosomes of* Drosophila melanogaster.*"*

his arms high and wide, exclaim "God, what a life!," and then bend back to his task. For anyone who knew Calvin Bridges and his love for fun and games, his unremitting toil was an unforgettable demonstration of scientific devotion. Bridges's chromosome maps set a worldwide standard and were universally used in correlating the location of genes with specific bands of the giant chromosomes.

Bridges returned to Cold Spring Harbor in 1938, but he suffered a heart attack during the summer. He was taken to the hospital and seemed to recover, but later that fall, back in Pasadena, he had a second, more serious attack. Within a couple of weeks, he was dead.

But his visit to Cold Spring Harbor in the summer of 1934 bore fruit; the first issue of the *Drosophila Information Service* came out that year.

Cold Spring Harbor Becomes a Resource for Drosophila *Genetics*

The *DIS* was a great success. From its initial run of 200 copies, it rapidly expanded in circulation and quickly became indispensable to *Drosophila* geneticists worldwide. Each edition included contributions from members: new stocks, lab tips for culturing or anesthetizing flies, and technical developments. Short abstracts of important new findings and information about technical matters too small for regular publication as a paper in some standard journal of genetics were also accepted.

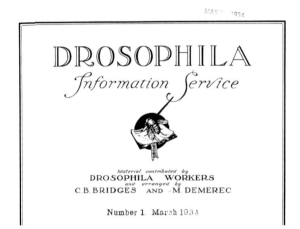

The cover of the first issue of the **Drosophila Information Service** *(DIS no.1-1934).*

The contents of the first issues exemplify the role *DIS* came to play in the *Drosophila* community. S. Gershenson contributed three research items on producing mutations using ether, on the distribution of X-chromosome lethals, and on the occurrence of a gene re-inversion, while A.H. Sturtevant and George W. Beadle discussed the confusion arising with the naming of X-chromosome inversions. Problems of nomenclature in general were a continuing theme of the *DIS,* with Muller contributing no fewer than four notes on nomenclature to issue Two. Methods for etherizing flies were also a recurring topic. Curt Stern described his apparatus in issue One, followed by both Demerec and Muller in issue Two. Early contributors included J. Schultz, C.B. Bridges, N. Timofeev-Ressovsky, L.C. Dunn, and B.P. Kaufmann.

Not all contributions dealt with research or technical matters. In issue Two, Muller raised the sensitive issue of how to acknowledge the source of materials presented in the *DIS.* Bridges and Demerec had stipulated that permission had to be sought from the author of the contributed note before it could be used in a publication, but Muller thought that this was

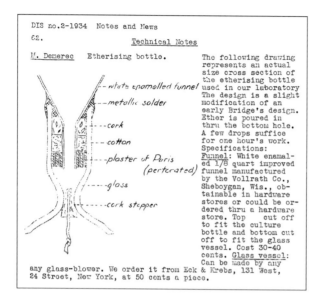

```
DIS no.2-1934  Notes and News
62.
                    Technical Notes

M. Demerec    Etherising bottle.        The following drawing
                                        represents an actual
                                        size cross section of
          --white enamelled funnel      the etherising bottle
                                        used in our laboratory
          --metallic solder             The design is a slight
                                        modification of an
          --cork                        early Bridge's design.
                                        Ether is poured in
          --cotton                      thru the bottom hole.
                                        A few drops suffice
          --plaster of Paris            for one hour's work.
                (perforated)            Specifications:
                                        Funnel: White enamal-
          --glass                       ed 1/8 quart improved
                                        funnel manufactured
          --cork stopper                by the Vollrath Co.,
                                        Sheboygan, Wis., ob-
                                        tainable in hardware
                                        stores or could be or-
                                        dered thru a hardware
                                        store. Top    cut off
                                        to fit the culture
                                        bottle and bottom cut
                                        off to fit the glass
                                        vessel. Cost 30-40
                                        cents. Glass vessel:
                                        Can be made by any
any glass-blower. We order it from Eck & Krebs, 131 West,
24 Street, New York, at 50 cents a piece.
```

Demerec submitted this sketch of the etherizing bottle, based on a design by Bridges, to the second issue of the DIS.

unnecessarily restrictive. The assumption must be made, he thought, that publication in the *DIS* showed that the author wanted the information to be used and that a simple acknowledgment was all that was necessary. This was a sensitive matter for Muller who had fallen out with Morgan, Sturtevant, and Bridges over what Muller perceived as a lack of appreciation of his contributions to the group. He concluded: "Experience has shown that it is unfortunately only too true that such overt understandings must be reached beforehand, even in the case of scientists, in order that real cooperation may be possible." Muller also contributed an item on a "Fly morgue" involving "used oil from automobiles."

Muller's piece was followed by a news item from Demerec. The drosophilists had a long tradition of supplying each other with strains carrying useful combinations of genes. Now with the help of the Carnegie Institution of Washington, the Rockefeller Foundation, and the American Society of Naturalists, a *Drosophila* stock center had been established at Cold Spring Harbor.

There were already stock centers for *Drosophila* at the California Institute of Technology and the University of Texas, but shipping stocks long distances could be risky to the flies and having an East Coast stock center would be a boon to drosophilists throughout the East. Demerec's initiative was not well received by T.H. Morgan, who feared Demerec was encroaching on his territory. Morgan cooled off after a while, though, and the stock center was established, greatly facilitating the work of East Coast drosophilists.

It is an indication of the status of Cold Spring Harbor in the *Drosophila* community that Demerec was able to bring two of Bridges's unfinished projects—completion of the salivary gland map and completion of the list of standard genetic symbols to be used in *Drosophila* papers—to the Department of Genetics. Morgan was again furious that Demerec was trying to take on work that he thought should be done at Caltech.

Demerec arranged for the completion of the salivary gland maps to be done by Bridges's son Philip, who had worked with his father during Bridges's last three summers at Cold Spring Harbor. Philip's Ph.D. advisor at Columbia, L.C. Dunn, agreed to allow the work to be included in Philip's dissertation, and Demerec arranged for Carnegie funds to support Philip as a Fellow of the Carnegie Institution. He worked on the salivary maps at Cold Spring Harbor in 1939 and 1940.

The symbol list also ended up at Cold Spring Harbor. Dunn recommended Katherine ("Kitty") Brehme as editor, and Brehme, then at Stanford, accepted the job, coming to Cold Spring Harbor for a 2-year appointment. She had just become engaged to Charles Warren, who had earned a Ph.D. under Eric Ponder at NYU. After their honeymoon, Brehme commenced work on what became *The Mutants of* Drosophila melanogaster, by C.B. Bridges (1944), completed and edited by K.S. Brehme. Kitty's role in editing some of the Symposia volumes was discussed in Chapter 7.

Katherine Brehme Warren (1942).

> # THE MUTANTS OF
> # DROSOPHILA MELANOGASTER
>
> CALVIN B. BRIDGES
> *Late Research Associate, Carnegie Institution*
> *of Washington*
>
> COMPLETED AND EDITED BY
>
> KATHERINE S. BREHME
> *Wellesley College*
> *Former Fellow of Carnegie Institution*
> *of Washington*

Cover of The Mutants of Drosophila melanogaster *(C.B. Bridges and K.S. Brehme, 1944).*

MacDowell Begins Research on Leukemia

Not all genetics at Cold Spring Harbor was done with flies. E. Carleton MacDowell continued his studies of mammalian genetics and cancer, the Carnegie Department's most medically oriented research program. In 1928, he began a collaboration with two physicians from the Department of Pathology at the New York College of Physicians and Surgeons to study lymphatic leukemia, a topic that occupied him for the rest of his career at Cold Spring Harbor.

MacDowell's main research tool was the mouse strain C58, developed by MacDowell in 1921. He found that lines of C58 varied in their susceptibility to inoculation with leukemia cells. By 1934, he had inbred lines of C58 in which 90% of individuals developed leukemia, pointing to a strong genetic component to this disease. When outcrossed, the heritability of leukemia dropped to 50%, and the backcross of these hybrid individuals with the non-leukemic strain reduced the heritability by half again.

MacDowell also became interested in immunological aspects of cancer. When he inoculated mice with very low numbers of leukemia cells, he found that he achieved "the active immunization of mice naturally 100 percent susceptible to a standard dose of highly virulent leukemic cells." Inoculation seemed to produce a resistance to leukemia similar to that found in naturally resistant strains. Although resistance to transplanted tumors could be induced, this resistance did not carry over to spontaneous tumors, and it had been thought that a special immune relationship existed between an animal and its own tumor cells. MacDowell's group, however, showed in 1936 that there is nothing special about a mouse's own leukemic cells; the failure to induce resistance in other studies was simply due to immunological differences between the two (transplanted and spontaneous) tumors. MacDowell later began to examine the nature of immunity to cancer, examining whether cell-free extracts of leukemic tissue could induce immunity. He determined that immunity, although inseparable from cells, did not require living cells; even leukemic tissue subjected to boiling heat could stimulate immunity.

MacDowell continued to work on leukemia at Cold Spring Harbor until his retirement in 1953, at the Carnegie Institution of Washington's mandatory age limit of 65, after 38 years of service to the Department of Genetics. His work on a heritable, transplantable form of leukemia in mice was a major contribution to early cancer research, undertaken at a time when few scientists were willing to tackle what seemed an intractable, mysterious disease.

Blakeslee Takes Up Human Genetics

In 1928, John Belling, who had worked with Blakeslee since 1920 (Chapter 4), moved to the University of California at Berkeley. Belling had suffered severe depression throughout his life, and, in 1925, matters were so serious that he was sent to Kings Park State Hospital, a large asylum for the mentally ill some 14 miles from Cold Spring Harbor. He remained there until his move to California—a move enabled by Davenport who continued to fund him.

E. Carleton MacDowell and Clare J. Lynch at the 1941 Symposium (IX, Genes and Chromosomes: Structure and Organization).

Dorothy Bergner replaced Belling as Blakeslee's principal coworker and they continued to work on chromosomal variation in *Datura* and its phenotypic consequences. Importantly, this diversity was found in "natural" plant populations—Blakeslee did not produce his trisomics by bombarding plants with X rays or other mutagens. In the 1930s, data such as Blakeslee's on mutations in natural populations formed the foundation for a major recasting of Darwinian evolutionary theory.

In 1937, Blakeslee and A.G. Avery discovered a chemical means of producing variations in chromosome number. They found that colchicine, extracted from the autumn crocus, paralyzed mitotic cell division and produced "polyploid" cells, in which the chromosome number was doubled, a discovery that Albert Levan described as "sensational." (Levan made an even more sensational discovery in 1956 when he and Joe Hin Tjio used colchicine in determining the correct number of human chromosomes.)

However, Blakeslee's most interesting work in this period led him into human genetics. In 1918, when Blakeslee and his assistant, B.T. Avery were examining verbena plants, Blakeslee noted that pale pink flowers were strongly fragrant, but that these flowers were without any scent for Avery. On the other hand, the red verbena flowers were fragrant to Avery but not to Blakeslee. They carried out a more systematic survey of 39 individuals and found that for 26 of them, the red flowers were fragrant while the pink flowers had no scent. The reverse was the case for the remaining 13.

Blakeslee took up the genetics of sensory perception some 13 years later when Arthur Fox, a chemist with DuPont, synthesized phenylthiocarbamide (PTC). His coworker complained of the extreme bitterness of PTC, but Fox insisted that it had no taste at all. Blakeslee read Fox's brief report and wrote to him asking for samples of PTC to test it more thoroughly. (The human geneticist Leonard Snyder at Ohio State University did the same.) Blakeslee reported his results in 1932, showing that all

	PARENTS		CHILDREN		
TYPE OF MATING	NO. OF MATINGS	NON-TASTERS (O)	TASTERS (T)		TOTAL

HEREDITY OF TASTE DEFICIENCY FOR PHENYL THIO CARBAMIDE
(Percentages in parentheses)

TYPE OF MATING	NO. OF MATINGS	NON-TASTERS (O)	CHILDREN TASTERS (T)	TOTAL
$O \times O$	10	22 (100)	0 (0)	22
$O \times T$	39	32 (43.2)	42 (56.7)	74
$T \times T$	54	22 (16.8)	109 (83.2)	131
Totals	103	76 (33.5)	151 (66.5)	227

Blakeslee's data show that all the children of non-tasters (O) are non-tasters, suggesting that they are homozygous recessives.

the children of two non-tasters were themselves non-tasters, whereas a taster child always had at least one taster parent. Tasting PTC was, then, inherited as a Mendelian trait, with "taste" being dominant to "non-taste."

Blakeslee's interest in the genetics of sensory perception continued for many years and at several meetings he set up stations at which participants could test themselves. At the Eugenics Conference in 1932, 6377 participants were tested for their responses to PTC, and at the International Flower Show in New York in 1935, more than 8400

Blakeslee with a test subject at the 1938 annual meeting of the American Association for the Advancement of Science.

attendees sampled a pair of flowers to determine whether they could detect an odor. Clearly, Blakeslee was fascinated with what his studies on taste revealed of the ways human beings interact with their environment. As human beings are born with different sensory levels, he wrote, so "… different people live in different worlds … so far as their sensory reactions are concerned."

Oscar Riddle and Hormones

Oscar Riddle continued to plow his lonely furrow as the only physiologist in the Department of Genetics. By focusing only on heredity and development, he said, his genetics colleagues took a two-sided attack on the many-sided problem of understanding hereditary characters. The better approach, he felt, would be to collaborate and address a class of problems from many angles. The class of problems that interested Riddle was reproductive physiology studied in the pigeon. He began these studies in the 1910s and published 33 papers over a 20-year period under the general title "Studies on the physiology of reproduction in birds." It was in the 30th in the series that Riddle reported the major discovery of his career.

STUDIES ON THE PHYSIOLOGY OF REPRODUCTION IN BIRDS

XXX. CONTROL OF THE SPECIAL SECRETION OF THE CROP-GLAND IN PIGEONS BY AN ANTERIOR PITUITARY HORMONE

OSCAR RIDDLE AND PELA FAY BRAUCHER

From Carnegie Institution of Washington, Station for Experimental Evolution, Cold Spring Harbor, N.Y.

Received for publication March 31, 1931

The anterior pituitary is the known source of two or three hormones whose separation and identification have not yet been accomplished in a completely satisfactory manner. Hormones from this gland have already been rather definitely associated with several different reactions or processes in the organism. Another wholly distinct reaction of an anterior pituitary hormone is described here.

Pigeons of both sexes brood or "incubate" their eggs. When young

Oscar Riddle and Pela Fay Braucher report their first findings using the pigeon crop as an assay for a hormone of the anterior pituitary.

The control of lactation was an enigma until 1928 when P. Stricker and F. Grueter found that extracts of the anterior pituitary induced lactation in rabbits, although they believed that prior ovulation was essential. George Corner, in work carried out at the Biological Laboratory during the summer of 1929, independently showed the same but demonstrated that ovulation was not necessary. He failed to isolate the hormone responsible. According to Corner, Riddle read Corner's report and decided to exploit the extraordinary phenomenon of "crop milk" secreted by doves and pigeons to feed their young.

It is not clear why Riddle chose to use the crop gland as an assay for a possible anterior pituitary hormone affecting lactation because, as he pointed out, the mammary gland and the crop gland are not homologous. Nevertheless, he and Pela Fay Braucher began to assay extracts of the anterior pituitary and other hormones for their ability to stimulate crop gland enlargement.

In the following year, Riddle made extracts of sheep and beef anterior pituitary that were relatively free of contaminants and were highly effective in stimulating crop gland enlargement. They named their new hormone "prolactin." That this was the hormone involved in mammalian lactation was demonstrated unequivocally when prolactin induced lactation in male and female guinea pigs and rabbits. In the Department's 1933–34 Report, Davenport reported that Riddle had provided prolactin to the Sloane Hospital of the Columbia University Medical Center for stimulating lactation in mothers and that prolactin was being produced commercially.

Working out the actions and interactions of the various pituitary hormones occupied Riddle for the rest of his scientific career. Although Riddle's section on endocrinology appeared anomalous among those on genes and chromosomes and mutations, it is clear that his research was a significant contribution to the work of the Department. He retired from the Carnegie Institution in October 1944 but not before completing two monographs summarizing his life's work on the metabolism and endocrinology of pigeons and doves.

"An Algebra for Theoretical Genetics"

In April, 1940, a Ph.D. thesis with this title was submitted to the Department of Mathematics at the Massachusetts Institute of Technology (MIT). It was the work of Claude Shannon, later called the "father" of information theory, and remarkably it was written in the summer of 1939 while Shannon was at Cold Spring Harbor.

Following graduation in 1936, Shannon had gone to MIT where he worked on the differential analyzer designed by Vannevar Bush. Shannon developed a rational approach to designing such systems in his 1937 Master's thesis, "A Symbolic Analysis of Relay and Switching Circuits" and is credited with laying the foundations of digital computer and digital circuit design theory. Bush was impressed by Shannon and wondered whether "… just as a special algebra had worked well in his

A section on random mating from Shannon's unpublished Ph.D. thesis.

hands on the theory of relays, another special algebra might conceivably handle some of the aspects of Mendelian heredity."

Vannevar Bush became President of the Carnegie Institution of Washington in 1938 and so it is not surprising that he sent Shannon to Barbara S. Burks, who was working in the Eugenics Records Office. She was carrying out careful surveys of the ERO records for traits and families that could be used for linkage analysis. In her studies of identical twins, she was insistent that both heredity and environmental variables had to be taken into account in analyses of pedigrees.

Shannon's approach to genetics was indeed theoretical: "... we shall speak ... as though genes actually exist and as though our simple representation of hereditary phenomena were really true, since so far as we are concerned, this might just as well be so." Shannon completed his thesis but never published the findings, despite much encouragement to do so. Even if Shannon's work had been published, it is unlikely that his theoretical approach would have found much favor with the leading mathematical geneticists—Lancelot Hogben, R.A. Fisher, Sewall Wright, and J.B.S. Haldane—whose work was firmly rooted in biology.

The Biological Laboratory

Eric Ponder, taking over as director in 1936, continued the Symposia, which rapidly became a fixture in the calendar, bringing excitement and a change of pace to the Laboratory. But on the research front things were far less exciting and year-round research was barely maintained. In any year during this period, only a handful of investigators were noted in the Biological Laboratory annual report as being "All-year" staff. The maximum number of investigators was seven in 1934 and 1935, but of these only four—Fricke, Harris, Grout, and Ponder—spent more than those 2 years at Cold Spring Harbor, and Grout's work on mosses was hardly

Eric Ponder.

in line with Harris's goal of a quantitative biology. Ponder was the only all-year round investigator after Fricke left in 1939.

Biophysics continued to be the in-house research program at the Biological Laboratory. Fricke's research seemed to become more and more physical rather than biophysical as he carried out experiments irradiating aqueous solutions with X rays. These resulted in papers with titles such as "The oxidation of the ferrocyanide, arsenite, and selenite ions by the irradiation of their aqueous solutions with x-rays." This work was regarded as important in its field but its relation to biology was obscure. Fricke did collaborate with Ponder, who continued to work on the membrane of the red blood cell.

The summer courses continued to attract students, enrolling a total of about 35 students in Plant Sociology, Marine and Freshwater

Zoology, Surgical Methods in Experimental Biology, and Experimental Endocrinology. Enrollment dropped, however, as the decade neared its close. In 1938, only the course in Marine and Freshwater Zoology was full, while just a few students enrolled in the courses in Experimental Surgery and Experimental Endocrinology and none at all in Plant Ecology. In 1940, total enrollment in the three courses offered was 12. The loss of income—$1700—was substantial. Clearly, the diminished

List of the investigators who worked at the Biological Laboratory during the summer of 1941. Many stayed on having been participants in the Symposium on Genes and Chromosomes: Structure and Organization.

INVESTIGATORS AND ASSISTANTS

Abramson, Harold A.—P. & S. and Mount Sinai Hospital, New York.
Anderson, Edgar—Missouri Botanical Garden, St. Louis, Mo.
Atwood, K. C.—Columbia University, New York City.
Austin, Robert—New York University, New York City.
Ayer, Richard B.—Choate Preparatory School.
Bartlett, James H.—University of Illinois, Urbana, Ill.
Bridges, Philip—Columbia University, New York City.
Carlson, J. Gordon—University of Alabama, University, Ala.
Christiansen, G.—Wellesley College, Wellesley, Mass.
Cordts, Edna—Vanderbilt University, Nashville, Tenn.
Corner, George W.—Carnegie Institution of Washington, Baltimore, Md.
Creighton, Harriet B.—Wellesley College, Wellesley, Mass.
Delbruck, M.—Vanderbilt University, Nashville, Tenn.
Diederich, Paul B. Mrs.—University of Chicago, Chicago, Ill.
Dunn, Jane—Cornell University Medical School, New York City.
Eigsti, O. J.—University of Oklahoma, Norman, Okla.
Ephrussi, B.—Johns Hopkins University, Baltimore, Md.
Geschicter, H.—University of Southern California, Los Angeles, Calif.
Giles, Norman H., Jr.—Yale University, New Haven, Conn.
Glass, H. Bentley—Goucher College, Baltimore, Md.
Gordon, Myron—New York Zoological Society, New York City.
Gurney, R. W.—University of Bristol, England.
Hoecker, Frank E.—University of Kansas City, Kansas City, Mo.
Hollaender, Alexander—National Institute of Health, Bethesda, Md.
Houlahan, Mary—Carnegie Institution of Washington, National Institute
    of Health, Bethesda, Md.
Hoyt, J. Southgate Y.—Cornell University, Ithaca, N. Y.
Jones, E. Elizabeth—Wellesley College, Wellesley, Mass.
Kimball, R. F.—The Johns Hopkins University, Baltimore, Md.
Kuchler, Frances—Cornell University Medical School, New York City.
Lewis, E. B.—California Institute of Technology, Pasadena, Calif.
Longacre, Dorothy—Fordham University, New York City.
Luria, Salvador E.—Columbia University, New York City.
McClintock, Barbara—University of Missouri, Columbia, Mo.
Mirsky, A. E.—Rockefeller Institute for Medical Research, New York City.
Morgan, D. T.—Columbia University, New York City.
Muller, H. J.—Amherst College, Amherst, Mass.
Nebel, B. R.—State Agricultural Experiment Station, Geneva, N. Y.
Nebel, Ruttle Mabel—Sate Agricultural Experiment Station, Geneva, N. Y.
Neel, James V.—Dartmouth College, Hanover, N. H.
Perrot, Max—University of Missouri, Columbia, Mo.
Rhoades, M. M.—Columbia University, New York City.
Schultz, Jack—California Institute of Technology, Pasadena, Calif.
Skoog, Folke—Johns Hopkins University, Baltimore, Md.
Sonnenblick, Benjamin P.—Queens College, Flushing, N. Y.
Vilkomerson, Hilda—Columbia University, New York City.
Warren, Charles O.—Cornell University Medical School, New York City.
Weaver, E.—University of Missouri, Columbia, Mo.
Wright, Sewall—University of Chicago, Chicago, Ill.

51

demand for such courses, particularly at the graduate level, required a reassessment of the Biological Laboratory's teaching program.

Yet, despite the paucity of year-round research and the falling numbers of course students, the Biological Laboratory annual reports were substantial. It was the summer investigators who sustained the Biological Laboratory as a research center and the "Reports of Investigators" filled the annual reports. The summer of 1941 was remarkable for the Symposium and the participants who stayed on to do research. The Symposium was on *Genes and Chromosomes: Structure and Organization,* and the 39 investigators included five who would later be awarded Nobel Prizes (Max Delbrück, Ed Lewis, Salvador Luria, Barbara McClintock, Hermann Muller, and Sewall Wright). Three of these would also help shape the future of Cold Spring Harbor. The Symposium topic and research topics of the summer investigators were harbingers of future research at Cold Spring Harbor.

Cold Spring Harbor at War

Demerec, in his annual report of 1941, wrote that that year marked the beginning of a third epoch in the Biological Laboratory's development. In the first, the Laboratory was created by the Brooklyn Institute of Arts and Science, and the second began when the Long Island Biological Association took over management of the Laboratory. The start of the third epoch was marked by Demerec becoming director on the resignation of Eric Ponder, so that the two institutions at Cold Spring Harbor were once again under the command of the same director.

Demerec's goal was to exploit the strengths of his two institutions, even if each remained under different administrative authorities—the Long Island Biological Association and the Carnegie Institution of Washington. The Biological Laboratory had land and buildings to support the summer research investigators and the Symposia, which by

Cold Spring Harbor (1947). Looking north along a rural Bungtown Road with Hooper Cottage on the right.

now were known throughout the world. The Carnegie Department of Genetics had a distinguished staff, a good library, and several buildings that housed its various laboratories. The two institutions thus had complementary resources that, combined, could create a powerful research and teaching facility. He moved swiftly.

The Biological Laboratory in Transition

The all-year-round research program of the Biological Laboratory had lived a hand-to-mouth existence for many years, and now the course program was severely diminished. Demerec decided to abandon both. For the foreseeable future, the Biological Laboratory would concentrate on the Symposium and the publication of its proceedings and on providing a summer home for researchers. The Carnegie Institution of Washington generously agreed to cover that portion of Demerec's salary that would otherwise have to be met by the LIBA. With other cost cutting, the Biological Laboratory ended 1941 with a small surplus.

Harris and Ponder had turned the Biological Laboratory to biophysics, a program now defunct with Ponder's departure. Demerec reintroduced genetics to the Biological Laboratory through the Symposium and the summer investigators. The 1940 Symposium on *Permeability and the Nature of the Cell Membrane* looked to the past; the 1941 Symposium on *Genes and Chromosomes: Structure and Organization* looked to the future. Although it would not be until after World War II that significant steps could be taken to answer the questions raised in the 1941 Symposium, that such a meeting could be held was an indicator of how genetics was to develop. That the timing was right was shown by the fact that although Demerec was expecting 35 to 50 participants, no fewer than 120 attended!

War Research at Cold Spring Harbor

Beginning in the fall of 1942, the Biological Laboratory became home to two research projects relating to the war. Demerec was approached by the

Airborne Instruments Laboratory (AIL) of the Division of War Research based at Columbia University. The AIL needed extra laboratory space, but why Cold Spring Harbor was attractive to them became known only later. Demerec cleared James biophysics laboratory and Urey, and later Cole Cottage, for the use of some 20 to 30 AIL scientists, engineers, and guards. Losing laboratory space was not the only inconvenience. The Biological Laboratory's machine shop was based in James, but because of the classified nature of the AIL work, the machine shop had to be moved out to a newly constructed building.

The nature of the AIL research was not learned until after the war when, at the request of the Laboratory's executive committee, Dr. O.W. Towner, director of the AIL, prepared a brief report and revealed why bucolic Cold Spring Harbor was attractive:

> The urgent need for an electrically quiet location for research and testing in the development of anti-submarine devices necessitated the establishment of a branch laboratory remote from the city. A survey of this area disclosed that your laboratory property was electrically and magnetically satisfactory … [the work at Cold Spring Harbor] undoubtedly shortened the time necessary to get the equipment into operational use… . Another development which resulted from our work at Cold Spring Harbor was a new technique and the necessary equipment for the measurement of the earth's magnetic field.

The second project was funded by a contract from the Chemical Warfare Service with additional funds from the Josiah Macy Jr. Foundation. It concerned the development of aerosols which "… would be found useful either in destroying germs and chemical substances while they were in the air or else in treating the lungs, by means of inhaled therapeutic agents, after lung injury had occurred." Harold Abramson was the leader of this work, with the assistance of Demerec, Vernon Bryson, and Bernice Samuels. They worked out a method for visualizing the size

Vernon Bryson relaxes on the tennis court, Davenport Laboratory in the background (1942).

of particles in various kinds of mists. A cheap and effective plastic nebulizer was also developed and put into commercial production. That was particularly important for therapeutic aerosols, because the size of the droplets had to be very fine if they were to reach the alveoli of the lungs. Bryson found that hydrogen peroxide aerosols were uniquely valuable in disinfecting rooms that were contaminated with bacteria adhering to dust particles. Even when the bacteria were embedded in the dust particles, very resistant microorganisms were thus destroyed.

A less successful project was to develop an aerosol nozzle that with low pressures would produce fine smokes for screening the movements of troops and mobile weapons. Such a nozzle was developed but proved less effective than other methods. However, it was effective in producing germicidal mists very rapidly, and Abramson in his report wrote that these nozzles were requested by one secret and two "top secret" installations during the war. Unfortunately, "the secrecy connected with the use of these atomizers was so great that the writer knows only partially the use to which they were put."

The Department of Genetics was less involved with war work but the Department did contribute in trying to alleviate a serious shortage of penicillin. Available techniques could not begin to produce enough of the drug to supply the front lines, let alone the general public, and a massive scaling-up of penicillin production was desperately needed. *Penicillium* grows either floating on the surface of the nutrient medium or submerged if the medium is sufficiently well oxygenated. The latter is more economical as regards medium, but floating cultures produce more penicillin.

Demerec had long been interested in mutations and how X rays and other forms of radiation induce mutations in *Drosophila*. Beginning in September 1943, he put this expertise to good use in trying to produce mutant strains of *Penicillium* that would produce large amounts of penicillin in submerged cultures. In May 1944, this work was put on a larger scale with funds from the War Production Board.

The project got under way with small-scale experiments using strain NRRL-832 before turning to the high-producing strain NRRL-1951. These experiments determined that treatment with 75,000 Roentgens was a suitable compromise between inducing mutations and killing the mold spores. The spores were irradiated at the Memorial Hospital in New York and spread out on agar plates back in the Department of Genetics. As soon as they had germinated, individual spores were inoculated into test tubes and cultures grown in shaking incubators. Samples were tested for penicillin production and those with high yields were saved and the rest discarded. It must have been hard work and rather depressing. They tested some 5000 mutants and found only 504 that were worthy of further analysis.

The Department did not have the facilities needed to test mold growth and penicillin yield under large-scale production conditions, so the promising cultures were tested elsewhere. They were shipped first to the University of Minnesota, where larger incubators were available, and mutants passing that test went to the University of Wisconsin, where they were tested in 80-gallon fermentation tanks. The end result was strain X-1612, which produced 300 milligrams of penicillin per liter, twice as much as the starting strain. A contemporary assessment of X-1612 was given by Robert Coghill, who had directed the penicillin efforts at the Northern Regional Research Laboratory: "… X-1612 is now very widely used by the penicillin industry and is producing what formerly would have been considered fantastic yields."

Demerec filed a patent on X-1612, but neither Demerec nor the Carnegie Institution of Washington benefited—the patent rights were assigned to "the United States of America, as represented by the Administrator, Civilian Production Administration."

Harry Warmke, who had joined Blakeslee in 1937, was responsible for an intriguing project. The Bureau of Plant Industry of the U.S. Department of Agriculture found itself overwhelmed with new plant-breeding projects

Patented July 27, 1948   **2,445,748**

UNITED STATES PATENT OFFICE

2,445,748

PRODUCTION OF PENICILLIN

Milislav Demerec, Cold Spring Harbor, N. Y., assignor to the United States of America as represented by the Administrator, Civilian Production Administration

No Drawing.   Application September 16, 1946,
Serial No. 697,380

2 Claims.   (Cl. 195—36)

This invention relates to the production of antibiotic substances, and more particularly to the production of penicillin.

It is an object of this invention to produce penicillin in extremely high yields. Another object is to produce mutations of molds of the genus mycetes capable of yielding extremely large amounts of antibiotic substances. Other objects will appear hereinafter as the ensuing description proceeds.

These objects are accomplished in accordance with this invention wherein an antibiotic-yielding organism of the genus mycetes is subjected to controlled dosage with ultra-short-wave radiation at an intensity insufficient to kill such an organism but amply sufficient to produce chromosome rupture therein whereby a mutant is produced, and thereafter such a mutant is cultured in a suitable propagation medium whereby a high yield of antibiotic substance is secured. Suitable mold organisms of the genus mycetes for use in accordance with this invention include *Penicillium notatum*, *Penicillium chrysogenum*, Actinomyces, and similar mold organisms of the genus mycetes. Preferably, there is employed an already high-yielding strain of *Penicillium chrysogenum*.

Suitable short wave irradiation for the production of mutants includes ultra-violet irradiation, cosmic irradiation, atomic fission irradiation, and preferably X-ray irradiation. For each of the above types of irradiation a dosage is chosen such that chromosome rupture is accomplished without killing the organism. In the case of X-ray irradiation a suitable dosage has been found to be between 10,000 and 150,000 roentgen units, and preferably there is employed a dosage of between 50,000 and 100,000 roentgen units. Under this dosage, the mutation is brought about in the spore form of the organism principally. In irradiating a culture containing numerous spores with the selected dosage of irradiation, it is usually found that the viability of many of the organisms is detrimentally affected. However, numerous of the remaining organisms are found, upon subsequent culture, to be mutants of the parent strain. Obviously, for commercial use of penicillin in antibiotic substances only the mutant or mutants producing the highest yields are propagated.

After irradiation, the mutant spores are spread on the surface of a suitable potato-dextrose-agar culture medium and allowed to germinate. Upon germination, the germinated organisms are isolated into the standard liquid lactose-corn steep liquor culture medium, made in accordance with the formula of the Northern Regional Research Laboratory at Peoria, Illinois, as described in the Journal of the Elisha Mitchell Scientific Society, vol. 61, page 78, for August, 1945. Thereupon, the cultures are agitated for a period of from 2 to 12 days. The resulting growth of the mold is treated in any suitable way for the recovery of its antibiotic content. For example, it may be extracted with amyl acetate, then treated with a small amount of activated carbon to purify the extract, and then further extracted with aqueous acetone. After separation from the activated carbon and the amyl acetate, the aqueous acetone solution is evaporated to yield an aqueous solution of penicillin or other antibiotic substance. Thereupon, by careful evaporation crystalline penicillin can be secured as a residue.

The following illustrative example shows how the invention may be carried out, but it is not limited thereto:

A seven-day-old culture of *Penicillium chrysogenum* (strain NRRL 1951.B25), grown in a test tube having the approximate dimensions, 1 x 7.5 cm., on standard potato-dextrose-agar medium, was subjected to X-ray radiation of 75,000 roentgen units at a rate between 2,000 and 3,000 units per minute. The X-ray irradiation was carried out with a usual d-therapy type of equipment commonly used for cancer treatment. After irradiation, spores were collected from the culture and spread on the surface of a potato-dextrose-agar culture medium contained in a Petri dish where they were allowed to germinate. Immediately after germination, the spores were isolated in separate test tubes containing 2 cc. each of liquid lactose-corn-steep medium made in accordance with the formula of the Northern Regional Research Laboratory at Peoria, Illinois. Thereupon, the test tubes were placed in a shaker machine having a four-inch horizontal stroke and operated at 250 strokes per minute. Each test tube had an approximate inside diameter of 10 mm. and a length when stoppered of 110 mm. The tubes were stoppered and the shaking machine operated for five days, after which the contents of each tube were diluted with 100 volumes of distilled water and assayed in duplicate for penicillin content by the Oxford cup method, using *Staphylococcus aureus* (NRRL strain B313). The tubes which showed a high yield of penicillin were saved for bulk fermentation tests and the low-yielding cultures were discarded. The following table shows the diameter of the inhibited region in millimeters for each Oxford

Patent 2,445,748 filed by Demerec on September 16, 1946.

and farmed out some of these to other institutions. One of the projects Warmke undertook as part of this program was to try to produce a line of fiber hemp with greatly reduced marijuana content. He reported that there was significant variation in cannabis content of extracts from individual

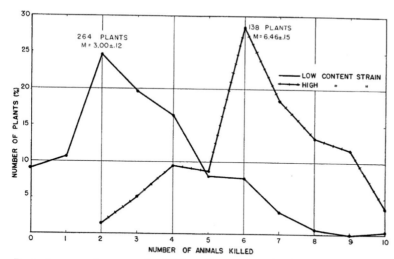

Fig. I. Average marihuana potencies of high- and low-content strains of hemp after three generations of selection, as measured by the *Daphnia* assay. The difference of the means is highly significant (Diff./S.E. = 18.21).

Graph showing the difference between the low- and high-cannabis-content hemp strains produced by Harry Warmke.

plants assayed by measuring how *Daphnia*, the water flea, was killed by an extract. Warmke was successful in selecting low-content strains and the following year was able to report that the low-cannabis strain had one-fourth the content of a high-producing strain. Samples were to be sent to the Bureau of Narcotics to determine whether "... the drug content may be approaching a level so low as to make it of little value to the vender of 'reefers'." Unfortunately the outcome is not recorded in the pages of the Annual Report.

Barbara McClintock

Demerec wrote in his 1944–1945 annual report that "[d]uring the war, because of the nature of our work, staff members of this Department were not called upon to participate in war research to any considerable extent." As a consequence research continued throughout the war with several important developments in staff and research projects, most notably the recruitment of Barbara McClintock.

The maize geneticist Marcus Rhoades was one of those who stayed on as a summer investigator after the 1941 symposium on *Genes and Chromosomes*. Rhoades had done his Ph.D. on maize with R.A. Emerson at Cornell University but was now at Columbia with no easy access to cornfields. Demerec, who had also worked with Emerson, willingly provided Rhoades with 5 acres of land for his use during the summer. Earlier in the year, Rhoades had learned that a former colleague of his from Cornell, Barbara McClintock, was having difficulties at the University of Missouri, where she had been since leaving Cornell in 1936. He prevailed upon Blakeslee to let McClintock share the 5 acres, on which he grew more than 35,000 and McClintock 7500 corn plants that summer.

Barbara McClintock accepts Blakeslee's offer of a place during the summer of 1941.

McClintock was already established as one of the finest cytogeneticists in the country. She had mapped the 10 known genetic linkage groups to each of the 10 maize chromosomes, establishing the basic link between cytology and genetics. In 1931, with Harriet Creighton, she showed that genetic and chromosomal crossing-over were the same event. In 1936, McClintock moved to Lewis Stadler's department at the University of Missouri, Columbia, to study the effects of X rays on maize. Using strains of X-irradiated maize provided by Lewis Stadler, McClintock demonstrated the existence of ring chromosomes, a new cytological phenomenon. This led to her seminal 1938 work on the breakage–fusion–bridge cycle, a repeating pattern of chromosomal recombination and instability.

However, McClintock's position at Missouri was primarily a teaching position, not a regular tenure-track faculty appointment. Stadler reassured her that it was as permanent as any other job, but McClintock had her doubts. She grew increasingly restless, felt Stadler was out to get her, and even talked about leaving genetics altogether. The pressure built until the spring of 1941 when she took an unpaid leave of absence but with no intention of returning.

Leslie Dunn and Barbara McClintock in conversation at the 1941 Symposium.

McClintock had a productive summer at Cold Spring Harbor in 1941, but at its end, she again faced the prospect of having no place to work. Demerec came to the rescue, arranging to pay McClintock through the end of the year and extended that for a further 9 months to September 1942 while he worked on a permanent solution. He set up an interview for her with Carnegie director Vannevar Bush, and wrote a glowing recommendation:

> if an opinion were asked of Dr. McClintock's colleagues they would rate her a first-class scientist and would place her among the first ten leading cytogeneticists of the world…. She has the reputation of doing her work so thoroughly and competently that there is no necessity for someone else to do additional work on the same problem…. She has a breadth of knowledge and vision that enables her to select problems of great significance; and, having planned the experiment, she works with considerable diligence and concentration.

```
                                              December 2, 1941.

        Dr. M. Demerec
        Department of Genetics
        Carnegie Institution of Washington
        Cold Spring Harbor, N.Y.

        Dear Doctor Demerec:
                I am replying to your letter of December 1st
        in which you extended to me on behalf of yourself and the
        Staff an invitation to join your group as a Guest Investigator
        for the month of December of this year.   I should be very
        delighted to join your group and wish to thank you and
        the Staff for their cordial cooperation and for the generous
        provisions which have been offered to me for continuation
        of my investigations.

                        Sincerely yours,

                        Barbara McClintock

                        Barbara McClintock.
```

McClintock's letter accepting an appointment as Guest Investigator in the Department of Genetics, December 2, 1941.

Demerec was successful and on April 1, 1942, Barbara McClintock became a staff member of the Carnegie Institution of Washington's Department of Genetics at Cold Spring Harbor. The Genetics Department staff were elated to have the illustrious young woman scientist as a colleague. When staff scientist E. Carleton MacDowell heard the news, he jumped in the air and said, "We should mark today's date with red letters in the Department calendar!"

The Biological Laboratory Acquires a Beach and a House

There was another reason to mark 1942 as a red-letter year. The Biological Laboratory had been acquiring parcels of land as they became available, both to provide room for expansion and to protect the beauty of the harbor. The Sand Spit dividing the outer and inner harbors is an essential part of the ecology of Cold Spring Harbor and the subject of Charles Davenport's first paper at the Biological Laboratory. However, it did not belong to the Laboratory but was part of an extensive estate belonging to Henry Wheeler de Forest and his wife who lived in their mansion Nethermuir. They gave express permission for resident and visiting scientists to use the beach on the north side of the Sand Spit, and there are many photographs of Symposia participants for swimming and picnicking there.

In 1942, de Forest's widow, Julia Noyes, demolished Nethermuir and disposed of the land but gave the Sand Spit to the Biological Laboratory. The Nethermuir site later became the CSHL President's House (see Chapter 17). The gift also included some 9 acres of land with a carriage house and stables. Demerec immediately used this building, appropriately known as the de Forest stables, for accommodations for scientists, a role it filled for more than 70 years until it became a child care center.

Nethermuir and the stables were not the only buildings on the de Forest estate. Airslie was built in 1806 by Major William Jones, a cousin of the John D. Jones who had provided the first land for the Biological

The Sand Spit is enjoyed by participants in the 1975 Symposium on **The Synapse.**

A view of Airslie in the 1960s.

Laboratory. In 1943 the Laboratory purchased Airslie and the surrounding acres and it became the Director's house.

Genetics and Evolution of Populations

Although research at Cold Spring Harbor was moving toward the newly developing genetics of microorganisms, studies of experimental evolution continued, albeit in a more sophisticated fashion. By the late 1930s, the modern evolutionary synthesis of Ernst Mayr, Sewall Wright, and Reginald Fisher had begun to provide new tools for understanding the genetics of populations and so to test directly Darwin's theory of evolution by natural selection, and to reconcile Mendelian heredity with Darwinian evolution.

In 1941, Demerec appointed Theodosius Dobzhansky, then at Columbia University, as a Research Associate of the Department of Genetics and included his work in the annual reports, perhaps because he felt that Dobzhansky's work fit so well with the intellectual tradition of Cold Spring Harbor. The work of Dobzhansky exemplified the new experimental approach to evolution studies. For example, in the

1943–1944 report, Dobzhansky described work as varied as examining the *Drosophila pseudoobscura* populations on Mount San Jacinto, California, by releasing and recapturing flies marked with a mutation such as orange eye color, laboratory studies of the effects of temperature on chromosomes of flies, collecting *Drosophila* in Brazil, and examining sexual isolation within a species. One of the founders of the new evolutionary synthesis, Sewall Wright, stayed on after the 1941 Symposium to help Dobzhansky analyze his data.

Another of the founders, Ernst Mayr, was a summer visitor every year between 1942 and 1952. Initially his reports included some experimental work on sexual isolation of species studied using *D. pseudoobscura* and *D. persimilis,* but increasingly he used the summer visits for writing papers and books. He completed the final editing of his classic *Birds of the Southwest Pacific: A Field Guide to the Birds of the Area between Samoa, New Caledonia, and Micronesia* at Cold Spring Harbor. As befits a great naturalist, Mayr also found time to examine the local flora and fauna, one year studying the insect fauna of mushrooms in Cold Spring Harbor, something of a contrast with his collecting expeditions in the tropical forests of Papua New Guinea.

Ernst Mayr on the Blackford Hall balcony during the summer of 1941.

A Footnote to the Eugenics Records Office

James Neel, then teaching at Dartmouth, first came to the Biological Laboratory in 1941 to pursue his research on mutations in *Drosophila,* but by the time he returned in 1942, he was thinking of taking up human genetics. That June, as he described in the Biological Laboratory Annual Report, he examined the files of the ERO to see if there was anything useful. He found a file labeled "red hair" with information of very varying quality on 26 families. The trait was thought to be recessive and, if so, all the offspring of redheaded parents should have red hair. Neel found 101 redheaded offspring of such parents, but there were also 13 with some other hair color, rather too many for a simple recessive trait.

Reprinted without change of paging from the Journal of He:edity (Organ of the American Genetic Association), Washington, D. C., Vol. XXXIV, No. 3, March, 1943.

CONCERNING THE INHERITANCE OF RED HAIR

JAMES V. NEEL

School of Medicine and Dentistry, The University of Rochester, Rochester, N. Y.

ALTHOUGH there is little doubt of the hereditary nature of red hair color, the details of the transmission of this trait remain somewhat obscure. It is generally agreed that the factors responsible for black hair are epistatic to those determining red. Even in dark brown haired individuals the detection of red pigment may be difficult. These facts, together with the probable existence of modifiers of the depth of red pigmentation, complicate the study of the inheritance of red hair color.

red heads are rare, only seven published ones being known to the author. The results of these former reports are summarized in Table I. From this it is obvious only that further data are needed.

Recently I have had the privilege of going through the files of the Genetics Record Office, Carnegie Institution of Washington, at Cold Spring Harbor, L. I., N. Y., and have been able to locate 26 additional reports of fertile marriages between two red headed individ-

James Neel's paper using records from the Eugenics Records Office.

Nevertheless, Neel published his paper, almost certainly the last paper based on data collected by Davenport and his legion of field workers.

The Nature and Action of Genic Substances

In 1944, Charles Davenport died. Although he had retired from the Laboratory in 1924 and from the Carnegie in 1934, Davenport remained on the boards of both institutions and continued to play what role he could in the institutions he had brought to prominence. According to Carleton MacDowell, in 1944 a whale washed up on the beach near the labs. Davenport, who had founded the Cold Spring Harbor Whaling Museum, rushed to claim the carcass and began cleaning it by boiling it. As ever, he did it with passion; MacDowell reports that Davenport "... became so permeated with the nauseating smell [of whale] that heads were turned when he approached ...". In the end, he worked not wisely but too well; he caught pneumonia to which he eventually succumbed. Whatever his faults, Davenport was a scientist till the end.

In the same year, Oscar Riddle retired, and Morris Steggerda moved to the Kennedy School of Missions, a division of the Hartford Seminary Foundation. The work of both men had become increasingly

Charles Davenport on Jones Beach in 1941.

anachronistic in the Department of Genetics, although Riddle made a lasting contribution to endocrinology and to Cold Spring Harbor. On the other hand, Steggerda had long ceased to work on topics relevant to the Department of Genetics; his latest work had been on anthropometry of Navajo Indians and anthropological studies of corn growing by Maya Indians, throwbacks to the studies he and Davenport had done 17 years earlier in *Race Crossing in Jamaica*.

In February 1946, Vannevar Bush, President of the Carnegie Institution of Washington, asked Demerec, Kaufmann, McClintock, and MacDowell to submit a proposed program of research. It is not surprising that they wrote, emphasis in the original:

> Thus we propose, as our primary objective, the elucidation of fundamental problems concerning the nature and action of genic substances.

The future direction of research at Cold Spring Harbor in the immediate postwar years and beyond was clear.

Morris Steggerda in 1939 on a field trip in Zuni, New Mexico.

```
Outline of a Proposed Program of Research

                    for the

          DEPARTMENT OF GENETICS
               Submitted by:


             M. Demerec
             B. P. Kaufmann
             B. McClintock
             E. C. MacDowell


        Cold Spring Harbor, New York
             February 23, 1946.
```

Title page of the memorandum prepared by Demerec et al. outlining the further development of the Department of Genetics.

BARBARA McCLINTOCK AND THE DYNAMIC GENOME

Summary

Barbara McClintock set cytogenetics on a firm foundation with a flurry of early discoveries. She devoted the remainder of her investigation to the identification, explanation, and significance of "controlling elements," later recognized as transposable elements, or more popularly "jumping genes," in the maize genome. Rather than the accepted static, orderly genome, she proposed one that was dynamic and changing—a radical leap in understanding.

The Discovery/The Research

In 1921, McClintock, then a student at Cornell University, enrolled in the only genetics course open to undergraduates even though she knew at that time that "genetics as a discipline had not yet received general acceptance." The cytology course with Lester W. Sharp focused on the structure of chromosomes and their behavior at mitosis and meiosis. To McClintock, "Chromosomes then became a source of fascination as they were known to be the bearers of 'heritable factors'." By graduation, she knew her advanced degree topic would involve "chromosomes and their genetic content and expressions, in short, cytogenetics."

Research results came quickly for McClintock with her distinguishing the 10 chromosomes of maize, and then with Harriet Creighton, showing genetic recombination is accomplished by physical rearrangement of the *Zea mays* chromosomes.

McClintock's work in the 1930s was in many ways preparation for the discoveries she would later make on transposable elements. McClintock found that structural changes (inversions, translocations, ring chromosomes) in maize chromosomes were induced by X-irradiation and involved breakage of the chromosomes. The ends from the same or different broken chromosomes were able to join together. Normal chromosomes did not fuse end-to-end, and broken chromosome ends were able to "heal," after which they could no longer fuse. Her insights into the ends of chromosomes has resonance with the 1990s research on telomeres and telomerase. During this period McClintock also discovered the nucleolar organizer region (NOR).

McClintock noticed that there were unusual patterns of pigmentation in kernels of plants with broken chromosome 9. In some maize families, this break occurred at the same site on chromosome 9; she named this locus *Ds* for dissociation. When *Ds* transposed along chromosome 9, it caused breaks, but it could only transpose in the presence of the second element, activator or *Ac*. McClintock called these mobile elements "controlling elements," which together represented a two-element system that controlled gene activity by turning genes on and off. In 1948–1950, she proposed that these movable genetic elements were heterochromatin and played a role in mutation, development, and evolution.

Barbara McClintock in her cornfield, 1953.

Continued

Barbara McClintock's corn.

In her now-famous 1951 Cold Spring Harbor Symposium presentation, McClintock described her work on *Ac* and *Ds* in maize in great detail. But McClintock's theory of genetic control of transposition and its role in development was resisted by the genetics community.

During the late 1950s and early 1960s, François Jacob and Jacques Monod proposed their model of genetic control in which a structural gene is flanked by controlling regions. McClintock was excited by these results and in 1961 published "Some parellels between gene control systems in maize and bacteria," in which she tried to align her *Ac* and *Ds* elements with the operator and regulator in the Jacob–Monod system.

A lesser-known aspect of McClintock's research was her work from 1962 to 1969 with the Agricultural Sciences Program of The Rockefeller Foundation. McClintock studied the geographic distribution of particular chromosomes and showed they revealed ancient migration and trade routes, thus deducing the origin of some species.

Significance

In the 1960s, further evidence of transposable elements began to be found in *Drosophila*, phage Mu, and bacteria. These simpler systems provided actual physical evidence for existence of insertion sequences and transposons as discrete DNA segments. And in the 1970s insertion sequences were recognized as transposable and the yeast cassette model of mating type switching was described.

Subsequent research has demonstrated the role of transposable elements/mobile genetic elements in antibiotic resistance in bacteria, production of antibodies, transformation of normal cells into tumor cells, and control of growth during development. The 1983 Nobel Committee awarded McClintock the Nobel Prize for Physiology or Medicine.

The Scientist

Born on June 16, 1909, in Hartford, Connecticut, McClintock had an unconventional childhood. She attended Cornell College of Agriculture earning her B.S. in 1923, M.A. in 1925, and Ph.D. in 1927. She remained at Cornell as an instructor from 1927 to 1931.

This time was followed by research fellowships at California Institute of Technology (she was the first female postdoc there) and in Germany during the rise of the Nazi regime, an experience she described as "not too happy." She returned to Cornell in 1934–1936 as a research associate and then moved to the University of Missouri in 1936–1941. She was invited to spend the summer of 1941 at Cold Spring Harbor by A.F. Blakeslee, and in December 1, 1941, Milislav Demerec asked her to come as to the Carnegie Institute's Department of Genetics as Guest Investigator; the next August she accepted a permanent position. She remained at Cold Spring Harbor for the rest of her life.

With her growing reputation, McClintock was elected in 1944 as first female president of the Genetics Society of America and was one of three women at that time to be admitted to the National Academy of Sciences.

McClintock received many awards for her groundbreaking research in addition to the Nobel Prize: the Presidential Medal of Freedom, MacArthur Prize Fellow Laureate, Albert Lasker Basic Medical Research Award in 1981, and the Wolf Foundation Prize. She died on September 2, 1992, in Huntington, New York.

— *J.C.*

The Postwar Transformation

Victory in Europe was declared in May 1945, but a return to normal conditions was not going to be achieved overnight. As Demerec wrote in the 1945 Biological Laboratory annual report, "Our expectations that the end of the war would allow us to return to a state bordering on normalcy have unfortunately not been fulfilled." It was particularly troubling in regard to the Biological Laboratory's infrastructure, which was in sore need of attention, and for Demerec's plans to relieve the chronic overcrowding in the summer by building new summer cottages. Such plans were abandoned because there was an acute shortage of materials and wartime restrictions on construction were still in place.

All was not gloomy. The Biological Laboratory had been able to obtain cheaply some expensive equipment surplus to government needs,

Revelers celebrate V-J Day, August 14, 1945, in Huntington, Cold Spring Harbor's nearby town.

and although the wartime penicillin project had been completed, the contract with the Medical Division of the Chemical Warfare Service still had 2 years to run. This enabled Vernon Bryson to continue his studies on the use of aerosols for therapies, particularly for the delivery of penicillin. As Robert Cushman Murphy, President of LIBA commented, LIBA members might rejoice that "... findings made under the auspices of the Chemical Warfare Service have turned out to be as important for the art of healing as for the ends of destruction."

But if Demerec was rather downcast by the present, his confidence in the future of Cold Spring Harbor was unshakeable.

Demerec's Manifesto

The memorandum that Demerec, Kaufmann, McClintock, and MacDowell sent to Vannevar Bush in early 1946 was a 68-page document that covered what they believed to be the future areas of exciting research, specific topics for the Department of Genetics, future recruitment, and the plans of the current investigators. They wrote that

[t]o maintain our leadership in this field [genetics], it is necessary to evaluate correctly the present state of the subject of genetics as a whole and to

```
                                                              To main-
tain our leadership in this field, it is necessary to evaluate correctly the
present state of the subject of genetics as a whole and to anticipate, as best
we can, the paths that will lead to the most enlightening contributions.  We
believe that the Department of Genetics should concern itself with the more
fundamental problems of heredity, the solutions of which will constitute the
center from which various ramifications stem.  Thus we propose, as our primary
objective, the elucidation of fundamental problems concerning the nature and
action of genic substances.
```

Demerec et al. set out the primary objective of future research in the Department of Genetics.

anticipate, as best we can, the paths that will lead to the most enlightening contributions.

The research program was divided into two broad categories: the nature of genic substances and the nature of genic action. Examples of the former included

An analysis of the physical and chemical nature of chromosomes

An analysis of the physical end chemical nature of the phage particle, a "naked gene"

Analyses of cytoplasmic substances having gene-like actions in microorganisms, in the mouse, and in maize

whereas genic action encompassed

An analysis of the changes in the action of phages that are associated with known changes in their constitution, and a consideration of the significance of such changes for our concepts of genic specificity and action

Determination in the mouse of the time of action, in early embryogeny, of single genic mutation and the type of cellular alteration that is produced to bring about a stream of highly modified cytoplasmic reactions during development

```
          If we consider the nature of genic substances, what are the problems
that face us?  The pertinent factors and questions revolve about the following
considerations:
     (1)  The chemical substances composing the gene.
     (2)  The structural organization of these substances.
     (3)  The nature of the reactive parts of the gene.
     (4)  The number of reactive centers in a single gene.
     (5)  The degree of diversity that single genes may show in composition,
structure, and reactive centers.
```

Part of Demerec et al.'s list of topics for study.

> An analysis of the cytoplasmic components in maize associated with instability of genic action, together with a consideration of the factors associated with their origin, their propagation, and their distribution to various cells during development

It is striking that although, as expected, microorganisms and *Drosophila* are listed for fundamental studies on the gene, so is the mouse, not least for an analysis of gene expression during mouse development!

Another farsighted proposal concerned interacting with the public: "Scientists ... should abandon the supercilious attitude that prevails in many quarters, and should make a planned attempt to satisfy the curiosity of the public." This should be done, not least, because as funding was going to come increasingly from public sources, "An enlightened public is more likely than an ignorant public to appreciate expenditure of its funds on scientific research."

A list was made of possible appointments to the Department of Genetics. Included were established scientists such as Delbrück, Luria, and Albert Claude, as well as rising stars like E.B. Lewis, but in the event, no one on the list came to Cold Spring Harbor.

Most interestingly, Appendix 3 dealt with the relationship between the Department of Genetics and the Biological Laboratory. It is clear that Demerec was upset that the opportunity had not been taken to develop a formal arrangement:

> It seems highly regrettable that a program for the Department of Genetics should have been prepared without full and frank discussion with an authorized joint planning board. ... The need is for a formal and legal contract stating the specific liabilities and responsibilities of the laboratories towards each other, including practical methods of accounting for time and funds, and, most especially, a method of harmonizing policies and programs.

Although Demerec was director of both institutes, the ultimate authority for determining the overall direction of the programs rested with the Trustees of the Carnegie Institution of Washington and the members of the LIBA. Demerec was particularly concerned that the LIBA had launched a membership drive based on the war work conducted in the Biological Laboratory. This indicated that the long-term interests of the LIBA were in "old" work, not in genetics. In the absence of a formal arrangement, Demerec suggested that a "full material and administrative separation of the two laboratories" should be considered. If Demerec hoped that this blunt account would stir his masters, it was not to be. Not until after his retirement, and the crisis precipitated by the Carnegie's withdrawal from Cold Spring Harbor, was there to be just one institution by the harbor.

Delbrück, Luria, and the Phage Course

The most important single event in the postwar years, an event that had a long-lasting impact on Cold Spring Harbor and genetics, was the founding of the Phage course by Max Delbrück and Salvador Luria.

Delbrück was a German theoretical physicist who moved to biology after having heard Niels Bohr's lecture *Light and Life* in 1932. Together with Nikolaj Timoféeff-Ressovsky and Karl Zimmer, he published a classic paper in which he derived an estimate for the size of the gene. In 1937, Delbrück came to the United States on a Rockefeller fellowship to learn genetics. He visited Cold Spring Harbor, but settled at Caltech, where he planned to work on *Drosophila* genetics. However, he soon began work with Emory Ellis, who was studying bacteriophage. Delbrück was "absolutely overwhelmed" by the potential of phage for genetic research. He moved to Vanderbilt University in 1940, and one year later came as a summer investigator to the Biological Laboratory.

Luria had been born in Italy and trained as a physician in Turin. In 1938, he fled from Mussolini's fascist regime to Paris, and then in 1940 from the Nazi invasion of France to the United States. He received a

Max Delbrück sits with the glants of modern physics at the Niels Bohr Institute, Copenhagen, Denmark, in 1933. From left to right in the front row: Niels Bohr, Paul Dirac, Werner Heisenberg, Paul Ehrenfest, Delbrück, and Lise Meitner.

Delbrück and Luria examining Petri dishes.

Rockefeller Foundation fellowship and began working on phage at Columbia University.

In the winter of 1940–1941, Delbrück and Luria met at the American Physical Society meeting in Philadelphia. They began talking science, and Delbrück proposed to Luria that they meet in Cold Spring Harbor for the Symposium and to carry out phage experiments. During the summer of 1941, they determined the lysis times of two types of phage and showed that when both types are present in a culture, the slower-lysing type suppressed the faster. This observation became known as their "mutual exclusion principle." Delbrück and Luria returned to Cold Spring Harbor in 1942 but the major contribution of Delbrück came after the war.

Delbrück developed a course on bacteriophage at Vanderbilt, but few students took it and those that did were not capable of the strict quantitative approach that Delbrück demanded. According to Delbrück, it was Luria who thought that "if ever phage was to become an important line of research, and its potentialities really developed, more people would have to

Students in the Phage course laboratory ca. 1945. The handwriting on the photo is that of Delbrück's wife, Manny.

be brought into it. And therefore one should make an effort to bring more people into it this way, by teaching the course." And so the first phage course was held at the Biological Laboratory, July 23–August 11, 1945.

Based on Delbrück's experience with the course at Vanderbilt, a prerequisite for the course was "[f]acility in the processes of multiplication and division of large numbers; elements of calculus; properties of exponential functions. … An admission test on these subjects will be given." Perhaps this is why only six students enrolled the first year! All, with one exception, were already established researchers. Two notable students were Rollin Hotchkiss and Herman M. Kalckar. Hotchkiss was a research associate with Oswald Avery at the Rockefeller and refined the classic transformation experiment to the point where protein contamination of the purified DNA was less than 0.02%. Kalckar, then at the Public Health Institute in New York City, moved his laboratory to the Institute for Cytophysiology in Copenhagen. It was there that Jim Watson did a postdoc with him, an experience Watson described as a "complete flop."

Curiously, if Delbrück remembered correctly that the course was Luria's idea, Luria never taught it. In the second year, it was taught by Delbrück again, when enrollment had doubled to 12 and hovered between 12 and 16 for the next several years. One of the 1946 class was Mark Adams, who returned the following year as a co-instructor with Delbrück. The following year and for the next seven years, Adams taught the course alone (with the help of assistants).

The phage course trained many of the next generation of phage workers; the roster for the first five years is a stellar collection of important biologists of the 1950s: Aaron Novick, Philip Morrison, and Leo Szilard (1947); Seymour Benzer, Seymour Wollman, Gunther Stent, and Bernard Davis (1948); Norton Zinder (1949); and Niccolo Visconti (1950). Especially notable are the physicists, including Benzer, Morrison, and Szilard, entering biology via the phage group. The students were treated to lectures by several future Nobel Laureates, including Delbrück, Luria, and Renato Dulbecco, as well as by many other soon-to-be-luminous alumni of the course.

The phage group continued to gather in the summers. In 1948, Luria came to Cold Spring Harbor, in part to collaborate with Demerec. He brought his postdoc, the Italian M.D., Renato Dulbecco, and his new first-year graduate student, James D. Watson. Watson fell in love with the "summer camp of science" at Cold Spring Harbor. He gave a charming description of in his essay, "Growing up in the phage group":

> As the summer passed on I liked Cold Spring Harbor more and more, both for its intrinsic beauty and for the honest ways in which good and bad science got sorted out. ... Most evenings we would stand in front of Blackford Hall or Hooper House hoping for some excitement, sometimes joking whether we would see Demerec going into an unused room to turn off an unnecessary light. Many times, when it became obvious that nothing unusual would happen, we would go into the village to drink beer at Neptune's Cave. On other evenings, we played baseball next to Barbara McClintock's cornfield, into which the ball all too often went.

It was not all work. One year, Delbrück dressed up as Ariel for a production of *A Midsummer Night's Dream* on the porch of Jones laboratory.

Max Delbrück as Theseus in a production of A Midsummer Night's Dream. *The stage is the porch of Jones Laboratory.*

The annual "graduation" at the end of the course was marked by a costume party and the imbibing of copious amounts of ethanol-fortified beverages.

Unverifiable but illustrative and entertaining legends sprouted about the brutal honesty of their interactions. When bored during a seminar, Delbrück would leave, walking straight up the center aisle of the darkened room, ensuring that his head intercepted the projector beam. Young Jim Watson, on the other hand, would remain in the room during a dull talk, but would begin crinkling through the pages of a newspaper.

Phage Course costume party (left) *and annual "graduation"* (right), *1952.*

A mythology grew up around the phage group, fostered by the members themselves, some of whom undoubtedly sensed the significance of their work. Nevertheless, students on the course were deadly serious about their work, and the phage course established a culture for molecular biology.

Research in the Biological Laboratory—The Discovery of Photoreactivation

The research staff of the Biological Laboratory until 1949 consisted of Vernon Bryson and Albert Kelner. When the war contract ended, Bryson turned to the induction of mutations in *Escherichia coli* by chemicals, although he later began using irradiation.

In the 1948 Annual Report, Demerec reported that staff scientist Kelner had made a "very significant discovery." Indeed, along with the work by Swingle on ACTH, it is arguably one of the two most important discoveries made by a scientist in the Biological Laboratory.

Kelner was a bacteriologist who had come to the Laboratory in 1946 to work on a project funded by Schenley Laboratories, a company manufacturing penicillin and other antibiotics. He used X rays and UV irradiation

Albert Kelner holding his son Robert, on the Sand Spit, Cold Spring Harbor, 1948.

Schenley Laboratories funded Kelner's research on antibiotic resistance.

to find mutants of *Actinomycetes* and other molds with increased synthesis of antibiotics. He later tried to induce antibiotic-producing mutants of bacteria and molds that did not normally produce antibiotics. This was a fruitless task; Kelner failed to find any antibiotic-producing strains in more than 78,590 cultures of irradiated *E. coli*. However, in January 1948, Kelner found an extraordinary range—more than 100,000-fold—in the degree to which UV-irradiated cultures of *Streptomyces griseus* recovered from the treatment. He thought that it might be temperature-related, but a series of careful experiments over a range of tightly controlled temperatures proved that this was not the case. The breakthrough clue came on September 4, 1948, when he wrote in his notebook:

> Noted that the 35° water bath was in the full light on the lab table. Noted too that in old exp. At room temperature, in which recovery was greatest, the spores were in transparent bottles on the Laboratory shelf exposed to day light.

Kelner had discovered the phenomenon of photoreactivation.

There is a twist to the story. Demerec allowed Kelner to stay on until May 1949 while Kelner began looking for jobs. On October 30, 1948, he

Table 2 from Kelner's first publication on photoreactivation. Streptomyces spores were irradiated with UV light and then exposed to light from two photoflood lamps and a projection lantern for varying lengths of time. There was a 1000-fold increase in survival after just 10 minutes exposure.

TABLE 2

EFFECT OF DURATION OF VISIBLE LIGHT ILLUMINATION ON RECOVERY

ILLUMINATION TIME, MIN.	VIABLE CELLS PER ML. OF SUSPENSION	RELATIVE INCREASE IN SURVIVAL RATE
0	2.5*	. . .
10	2.5×10^3	1,000-fold
20	9.2×10^3	3,700-fold
30	1.3×10^5	52,000-fold
40	1.6×10^5	64,000-fold
50	2.0×10^5	80,000-fold
60	5.3×10^5	210,000-fold
145	5.5×10^5	220,000-fold
173	7.7×10^5	310,000-fold
240	8.0×10^5	320,000-fold

* The count of the non-ultra-violet irradiated suspension was 4.2×10^6, so that the survival rate at time zero was 6.0×10^{-7}.

wrote to Luria describing his photoreactivation experiments, and asking if Luria knew anyone at NIH who might be interested in hiring him. Luria wrote back on November 26 that "… Dulbecco has discovered, quite by chance…" what appeared to be photoreactivation in phage. This was followed by another letter from Luria saying that Dulbecco was preparing a short note for *Nature* describing his results. Kelner must have been devastated—here he was trying to find a job and Dulbecco was about to scoop him on what might well prove to be the discovery of Kelner's life!

Kelner's reply was courteous and diplomatic but left no doubt about what he thought had happened:

> … I want to first explain to you as frankly as I can some of my personal reactions to your letters … it seemed a most unusual, and almost impossible-to-believe coincidence that Dulbecco's discovery should have entirely independently been made precisely 3–4 weeks after I had written you the essentials of my findings … I cannot help feeling … that my findings had influenced the discovery of phage photoreactivation … . I would have felt much better if my original discovery and its relation to Dulbecco's were mentioned in your ms. to *Nature*…

Luria's response was equally courteous and recognized that an injustice might be done to Kelner. He proposed the addition of a paragraph to the *Nature* note stating that Kelner had priority in that he had communicated with Luria prior to Dulbecco's discovery; and that Dulbecco and Luria would delay publication until after Kelner had published. It was a generous response and Kelner replied to Luria "with best wishes and thanking you and Dulbecco for your honest and sincere reaction to my letters." Kelner's paper was published in the February 15, 1949 issue of *Proceedings of the National Academy of Sciences,* and Dulbecco's, with the additional paragraph, in *Nature,* June 18, 1949.

However, despite Demerec's increasing passion for bacteria and the novelty of Kelner's findings, Demerec seems to have made no effort to keep

> The occurrence of photo-reactivation of ultra-violet irradiated phage was noticed accidentally a few weeks after receiving a personal communication from Dr. A. Kelner that he had discovered recovery of ultra-violet treated spores of Actinomycetes upon exposure to visible light. I am informed by Dr. Kelner that his results are in course of publication[3]. My observation indicates the correctness of Dr. Kelner's suggestion that the phenomenon discovered by him may be of general occurrence for a number of biological objects.

The paragraph in Dulbecco's letter to Nature, *June 18, 1949, recognizing Kelner's priority in the discovery of photoreactivation.*

him. Kelner went first to Harvard University as a U.S. Public Health fellow and then in 1951 moved to Brandeis as an assistant professor of biology.

The Department of Genetics—The Nature of the Hereditary Materials

The research of the Department of Genetics in the postwar period is best summed up by the sentence with which Demerec opened his 1949–1950 report to the Carnegie Institution of Washington: "The focus of interest continued to be the nature of the hereditary materials, genes and chromosomes, using a variety of approaches and several different organisms in their experiments." Demerec also wrote that "The pattern of research … did not change appreciably during the past year" and in large part this was true of this postwar period as a whole.

McClintock continued her studies of mutable loci in corn and by the end of the decade had determined that two loci, *Ds* and *Ac,* were involved. Her interpretation of the phenomenon she was observing was that *Ds* and *Ac* chromatin could transpose (move) in the genome and if they became inserted next to an active gene, that gene could be turned off:

> A normal, wild-type locus may be totally or partially inhibited in action by the insertion of a foreign piece of chromatin adjacent to it. Total or partial release from inhibition will occur when this foreign chromatin is

removed or altered in organization. The insertion, removal, or change in organization of the foreign chromatin occurs because this chromatin becomes adhesive in certain somatic cells at very precise times in the development of a tissue.

It seems from Demerec's summaries of McClintock's work in the annual reports that the idea that pieces of a chromosome could move was not heretical. Demerec did not comment on that aspect of McClintock's findings, and it was not McClintock's major interest. As she wrote in her section of the 1949–1950 annual report, the behavior of these transposable elements "… may reflect one aspect of its normal behavior in the development of an organism."

In contrast to the complex genetic system studied by McClintock, Demerec was now a confirmed bacterial and phage geneticist. With visitors to the department, including Luria who spent a sabbatical year in his laboratory, Demerec's research focused on the nature of bacterial resistance to antibiotics, radiation, and phage infection. Each of these external insults to the genome was capable of provoking spontaneous resistance in a small subset of a population; thus, they provided a way of examining the nature of genetic change and the nature of the gene.

Kaufmann's studies of the organization of chromosomes had become increasingly visual, focused on cytology, including localization of cellular organelles, and chemical, examining the chemistry of cellular enzymes. MacDowell continued his experiments on mice resistant to a transplanted leukemia. There was a danger that his work of many years might have been compromised by the presence of a previously undetected virus. However, he still obtained the same results with virus-free cells.

By the end of 1949, Demerec must have felt pleased with the state of research at Cold Spring Harbor. The Biological Laboratory had

Expense:		
Symposia:		
Publication of annual Symposia on Quantitative Biology	$11,153.82	
Expense of participants and lecturers	3,776.96	$14,930.78
Dining Hall		10,151.36
Rooms and apartments		1,924.85
Research expenses		1,768.47
Summer course expense		480.85
Distribution of John D. Jones Scholarship		250.00
Buildings and grounds maintenance:		
Salaries	$ 5,475.55	
Materials and Supplies	5,516.55	
Heat, light and water	1,837.22	12,829.32
General and Administrative:		
Salaries	$ 3,365.90	
Insurance	599.86	
Printing and stationery	614.33	
Telephone, telegraph and postage	332.82	
Miscellaneous	980.90	5,893.81
Total expense		48,229.44
		$ 4,915.16
Provision for reserve for scientific research		3,000.00
Excess of income over expense		$ 1,915.16

Biological Laboratory budget showing a small surplus.

ended its 60th year in better shape than it had been for many years. The Department of Genetics was active and well supported by the Carnegie Institution of Washington. The Symposium had not suffered from the hiatus during the war years, and it had quickly reestablished itself as the place for communicating the latest research in the new and rapidly expanding field of molecular genetics. The future looked bright.

Molecular Genetics Comes of Age

The diamond anniversary of the founding of the Biological Laboratory opened on an optimistic note for both the Biological Laboratory and the Department of Genetics. Although the two institutes remained under different management, Demerec as director of both was promoting a greater synergy in their goals. His reports to the Carnegie Institution of Washington now included a description of the interactions between the staffs of the Biological Laboratory and the Department of Genetics.

Financially, the Biological Laboratory was doing well. In 1950, the income of the Biological Laboratory for the first time exceeded $50,000 and with Demerec's careful housekeeping, there was an excess of income over expenditure of $1,099. (Demerec was known for wandering the grounds at night turning off lights in unoccupied rooms.) Attendance at the Symposia was strong as were sales of the Symposia volumes; money from the latter was a substantial fraction of the Biological Laboratory's income.

With Carleton McDowell's retirement in 1952, the last link with Experimental Evolution was gone. McDowell had spent a lifetime at Cold Spring Harbor, coming first as a student in summer courses, 1908–1910, and then returning as a staff investigator in 1914. He must have felt rather isolated by the end of his career, surrounded by phage and bacterial geneticists. Microbial genetics was where the excitement was, and Cold Spring Harbor was in the vanguard.

Milislav Demerec in the 1950s.

Demerec Laboratory and Bush Auditorium

The optimism of the time is best demonstrated by two new building projects. The Department of Genetics was still based in the building that had housed the Station for Experimental Evolution. Recognizing the inadequacies of that building for modern research, the Carnegie Institution of Washington committed to building a new laboratory. At long last the building of the original Eugenics Records Office proved useful—Demerec sold it and the surrounding land to help finance the construction of the 16,000 square foot, state-of-the-art laboratory. The SEE building was to be converted into a library.

The Biological Laboratory's Symposia had long outgrown the Assembly Room in Blackford Hall, so in addition to the new laboratory, the Carnegie Institution of Washington also provided $100,000 to build a new assembly room. Built by the same architects as the new laboratory, and adjacent to it, the lecture hall was state-of-the-art. It had no parallel surfaces and the rear wall was a baffle, all to enhance the sound of

The new laboratory (later named for Demerec) under construction in November, 1951, with the original Carnegie building in the distance.

*Demerec Laboratory nearing completion
in October, 1952.*

speakers. There was a chalkboard stretching the width of the room and
seating for 250 meeting participants.

Construction was delayed because of a tremendous increase in building costs (50% according to Demerec) associated with the Korean War but
got under way in August 1951. The lecture hall and laboratory were formally declared open by Vannevar Bush on May 29, 1953. Bush remarked
that "… there is no science that is more fascinating, more intriguing, than

*Bush Lecture Hall approaching
completion in 1952, with Blackford Hall
beyond.*

the science of biology as it today flourishes," and that if he were a young man again, this was the field into which he would plunge. But it was the beginning of the Cold War, and Bush tempered his optimism with the fear that "… all of these efforts may be terminated in a struggle of desperate nations…."

The Symposia

The new lecture hall—later to be named for Vannevar Bush—was completed just in time for the 1953 Symposium on *Viruses*. Organized by Max Delbrück, it was held at just the right moment—the list of participants was a hall of fame of early molecular genetics. In his introductory remarks, Delbrück wrote that, "Special mention should be made of a last minute addition to the program, or rather, to the list of participants." The addition was the 25-year old Jim Watson. Watson had told Delbrück, one of his mentors, of the discovery of the double helix in a letter dated March 12, 1953, and Delbrück distributed copies of the Watson and Crick *Nature* paper to participants.

Bush Lecture Hall in use for the 1965 Symposium on Sensory Receptors.

Norton Zinder, just a few months younger than Watson and a graduate student with Joshua Lederberg, was a participant. Zinder had discovered transduction, in which bacteriophage carry DNA derived from one bacterial cell to a second. It was an exciting finding and as Zinder wrote, he had hoped to be one of the stars of the Symposium. "Though my paper was well received, I was disappointed by being overshadowed by the Watson–Crick model of DNA."

Although what later became known as molecular genetics was the cutting edge of genetics research, Demerec did not neglect other areas of biology in his selection of topics for the Symposia. Throughout the 1950s, he alternated topics in molecular genetics with a broad range of topics, some of which seemed out of place at Cold Spring Harbor. The 1950 and 1955 Symposia on the *Origin and Evolution of Man* and *Population Genetics: The Nature and Causes of Genetic Variability in Population* were at least connected with genetics, but *The Neuron* (1952), *The Mammalian Fetus: Physiological Aspects of Development* (1954), and *Population Studies: Animal Ecology and Demography* (1957) were far afield. Of these three disciplines, only neuroscience reappeared as a Symposium topic, although not for 13 years!

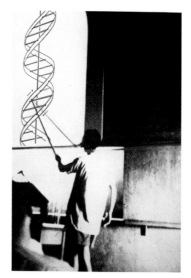

Jim Watson talks about the DNA double helix at the 1953 Symposium.

The Meetings and Courses

It is not surprising, given the success of the Phage course, that the second meeting added to the Biological Laboratory's year was on phage genetics. Phage researchers had been holding small meetings since 1947, but given the coincidence in 1950 that several phage workers were at Cold Spring Harbor in August, a two-day meeting was held, August 21–22, 1950. It was attended by about 35 of the leading phage researchers, including Jim Watson, Mark Adams, Al Hershey, Max Delbrück, Al Garen, Guiseppe Bertani, Gus Doermann, Seymour Cohen, Angus Graham, Aaron Novick, Tom Anderson, and Salvador Luria. As usual there was time to relax. Demerec described in the annual report that:

A square dance was held during the later part of the first evening, and a beer-and-pizza party concluded the proceedings of the second night. In the court of this last session M. Delbruck was tried in the Yokel Court of Long Island on charges of spitting into his cultures. Judge Loevinger [radiation biologist] presided. After numerous witnesses had been called and cross-examined the jury returned a verdict of "guilty", and the defendant was sentenced to twenty years of hard labor in the California Penitentiary of Technology.

The meeting was held again in 1951, but it was not until 1959 that it returned and became a permanent fixture at Cold Spring Harbor. The meetings program at Cold Spring Harbor arose from these humble beginnings, albeit slowly—on average from 1950 to 1968 there were fewer than two meetings each year. This is rather surprising given the success, intellectual and financial, of the Symposia; perhaps Demerec felt that the Biological Laboratory did not have the facilities to cope with more than this. The major change came when Jim Watson became director, but even then it was not until 1977 that double figures were reached.

There was also a doubling of the course program from one to two courses with the addition of a course on Bacterial Genetics. Taught by Demerec, Evelyn Witkin, and Vernon Bryson, the course enrolled 12 students, including Alan Garen, Niccolo Visconti, and Marguerite Vogt, all three of whom went on to make important contributions to science. Notable contributors to the course included Rollin Hotchkiss, who joined Oswald Avery's Rockefeller lab after Maclyn McCarty left in 1946. He demonstrated methods of studying transformation in *Pneumococcus*, and lectures were given by Max Delbrück, MacFarlane Burnet of Australia, and Mark Adams. The two courses continued until 1971, when the Phage Course ended after a quarter of a century. The Bacterial Genetics course (under different names) has passed its diamond jubilee.

A collage of photographs assembled and annotated by Manny Delbrück from the party at the end of the 1951 Phage course. Max Delbrück is standing, wearing the robe with white collar.

Hershey and Chase—Proving That DNA Is the Genic Substance

The Phage course and the phage researchers that came each summer had made Cold Spring Harbor a central player in phage genetics, but Gus Doermann was its only full-time phage scientist. That changed with Demerec's appointment of Alfred Hershey in August 1950. Hershey's career began in the Department of Bacteriology & Immunology at Washington University, where he worked with Jacques Bronfenbrenner examining factors affecting bacteria growth in culture. He published an

Leo Szilard and Alfred Hershey on a rainy day in Cold Spring Harbor in 1951.

early paper on phage in 1941 ("The influence of host resistance on virus infectivity as exemplified with bacteriophage"), but it was apparently a visit to Delbrück at Vanderbilt in 1943 that turned Hershey's attention full-time to phage. By 1950, he was a leader in the field of phage genetics, and Demerec must have been thrilled to have him at Cold Spring Harbor.

Hershey was not the only August arrival at the Department of Genetics. In the spring of 1950, Demerec received a letter from a young woman named Martha Chase at Wooster College in Ohio. She was about to graduate with a degree in biology and wondered if Demerec might have a research assistantship. After writing to her biology professor and receiving a positive letter of recommendation, Demerec replied he did have a position and she would be welcome to it if she liked. Demerec ended up giving the position to someone else, and so he asked Hershey if he would like a research assistant. Hershey, still at Washington University, took her sight unseen, replying, "I should be very glad to have Miss Chase if she decides to accept." Chase hesitated before accepting the position—her interest was bacterial genetics, not phage—but agreed to begin on August 15, two weeks after Hershey's arrival.

Hershey was already interested in the fate of phage DNA while at Washington University. He and his colleagues had labeled DNA of phage

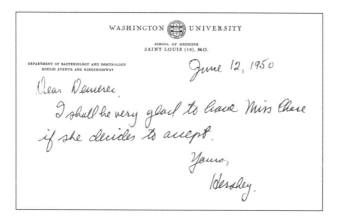

WASHINGTON UNIVERSITY

SCHOOL OF MEDICINE
SAINT LOUIS (10), MO.

DEPARTMENT OF BACTERIOLOGY AND IMMUNOLOGY
EUCLID AVENUE AND KINGSHIGHWAY

June 12, 1950

Dear Demerec,

I shall be very glad to have Miss Chase if she decides to accept.

Yours,

Hershey.

Hershey writes to Demerec telling him that Hershey will be glad to have Martha Chase in his laboratory.

with ^{32}P to "… determine whether all or only one of the progeny of a labeled phage particle contains isotopic phosphorus." His first report for the Department of Genetics began with a description of his first steps to label phage DNA and protein with ^{32}P and ^{35}S, respectively, and follow "… the transfer of phosphorus and sulfur from viral parent to progeny."

The following year's report, by Hershey and Chase, described a significant advance inspired by T.F. Anderson's electron micrographs showing that phage particles after infecting a bacterial cell remained attached to the bacteria by their thin "tails." This suggested that if these "ghosts" could be stripped from the bacteria, it would be possible to separate whatever phage materials had entered the cell from those that did not. It was apparently Margaret McDonald who lent Hershey and Chase a new appliance becoming popular in those days of household technologies and cocktail parties: a Waring Blendor. The blender provided just the right shearing action to rip the phage ghosts off the bacteria without rupturing the fragile bacterial cell membranes. The preliminary results suggested that

the viral nucleic acid functions as the sole agent of genetic continuity in the virus, and that the synthesis of viral protein (likewise a specific substance

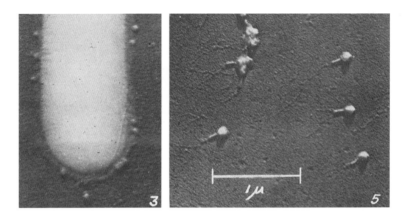

An electron micrograph from T.F. Anderson's presentation at the 1946 Symposium showing (left) *phage particles attached to a bacterial cell and* (right) *the head and tail morphology of the phage.*

foreign to the uninfected cell) occurs de novo as a late step in viral growth. These ideas may prove correct, incorrect, or untestable, but they are likely to orient research on viral growth for some time to come.

These ideas were testable, and the resulting paper "Independent Functions of Viral Protein and Nucleic Acid in Growth of Bacteriophage" was an instant classic. It was found that up to 80% of the ^{35}S, but less than

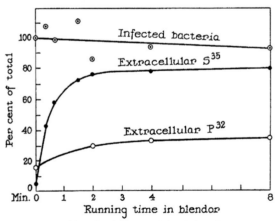

A graph from the Hershey–Chase paper showing that 80% of phage protein (labeled with S^{35}) but only 20% of phage DNA (labeled with P^{32}) can be stripped from the cells using the Waring blender.

FIG. 1. Removal of S^{35} and P^{32} from bacteria infected with radioactive phage, and survival of the infected bacteria, during agitation in a Waring blendor.

25% of the ^{32}P was stripped off. Thus, ~75% of the DNA had demonstrably entered the cells whereas no more than 20% of the protein had done so. And as none of the protein label appeared in the progeny phage particles, but about 30% of the DNA label did, Hershey and Chase concluded that the hereditary material of the phage was probably DNA.

It is striking that Hershey and Chase's conclusions about the role of phage DNA were as tentative as those of Avery, MacLeod, and McCarty on the role of DNA in pneumococcal transformation. Hershey and Chase wrote, "The chemical nature of the genetic part must wait, however, until some of the questions asked above have been answered." And although posterity has linked the blender experiments of 1951 with the transformation experiments of 1944, Avery was not cited by Hershey and Chase. It seems probable that the blender experiment was part of Hershey's quest to elucidate the process of phage replication, rather than to determine the chemical nature of the genetic material.

After 1952, a string of bright young scientists came to work with Hershey, including Alan Garen, Elizabeth Burgi, Norman Melechen,

Hershey Laboratory in mid-1950s
(left to right): N. Visconti, M. Chase,
A. Hershey, C. Chadwick, N. Symonds,
J. Dixon, and A. Garen.

and Niccolo Visconti. Hershey's group further explored the role of DNA in phage inheritance, but toward the end of the decade Hershey's interests to turned to biophysical studies of phage DNA.

Evelyn Witkin Glimpses a Second DNA Repair Mechanism

Evelyn Witkin joined the staff of the Department of Genetics in 1950, but she first came to Cold Spring Harbor when she was doing a Ph.D. with Dobzhansky at Columbia University. She intended to study chemically induced mutations in *Drosophila* but on reading the classic Luria–Delbrück paper on mutations in *Escherichia coli,* decided that bacterial genetics was the way to go. Dobzhansky proposed that Witkin spend the summer of 1944 working on *E. coli* in Demerec's Department of Genetics. Demerec suggested that she try to induce mutants of *E. coli* using UV light, but on her first attempt she killed all the cells except four. These she showed were 100-fold more resistant to UV than wild-type cells. Dobzhansky did not have suitable facilities at Columbia for

Evelyn Witkin and Gus Doermann outside Blackford Hall in the early 1950s, when parking on Bungtown Road was allowed (annotation by Esther Lederberg).

her to continue this research, and so Witkin returned to Cold Spring Harbor to complete her thesis work. She stayed on as a Carnegie fellow continuing to study one of the resistant strains, B/r; these studies led to the recognition of a pathway for DNA repair of UV-induced mutations.

Demerec and Elise Cahn *also* worked on B/r. They exposed *E. coli* to UV and measured the appearance of bacteriophage-resistant mutants at various times after exposure. They correlated the continuing appearance of mutants with the number of divisions cells underwent following irradiation. They controlled the number of divisions by varying the amount of the nutrient broth; the more broth, the more cell divisions and the greater the number of postirradiation mutants.

Witkin realized, however, that although the delayed appearance of mutants was correlated with the number of postirradiation cell divisions, it might be a consequence of the richness of the culture medium, rather than of the number of cell divisions. She found that exposure of cells to a rich medium for only 60–90 minutes following irradiation led to high levels of mutants, no matter how the cultures were subsequently treated; it was the medium present in the first hour that determined the final yield of mutants. Witkin found that adding a pool of amino acids to the minimal medium was as effective as complete medium. These findings suggested that protein synthesis was important. This was confirmed when chloramphenicol, an inhibitor of protein synthesis, was added to the cells during the first hour of postirradiation culture and the yield of mutants fell dramatically.

Witkin's interpretation of what came to be called mutation frequency decline (MFD) was, with hindsight, straightforward. She suggested that there were two processes at work during the "sensitive" period following irradiation: DNA replication and a process "X," which she believed "… is the process of repair of genetic damage." She continued, "… if repair is accomplished before the end of the sensitive period (DNA duplication?) the damaged cells survive, with a high probability of surviving as a mutant. If repair is not accomplished within the critical period, the damage is lethal." This repair process is light independent and is known as "dark repair," in contrast to photoreactivation discovered by Kelner.

Vannevar Bush visited Cold Spring Harbor in early 1949, and Witkin relates that she asked him if she could take a leave of absence after the birth of her first child:

> His reaction astounded me. He made a brief, impassioned speech about the unfairness of the difficulties facing women who want to have children and careers in science and declared that "institutions have to change to make it work." He then asked me what I would need to make it work for me. Scarcely able to believe what I was hearing, I told him that, ideally, I would like to come back to work on a part-time basis for a number of years. "Done!" he said. And that's the way it was until I left Cold Spring Harbor in 1955.

Witkin left the Department of Genetics for the State University of New York College of Medicine but returned to present her findings at the 1956 Symposium on *Genetic Mechanisms: Structure and Function.*

Demerec—Father and Daughter—Identify "Assembly Line Genes"

For more than 18 years, Demerec's research reports to the Carnegie Institution of Washington had appeared under a section titled "The Gene," but in 1950 he changed the title to "Bacterial Genetics," reflecting the research that would occupy him throughout the 1950s.

He focused on studies of mutations—how to induce them and the nature of resistance to them—initially using on *E. coli.* However, his most significant research was carried out using *Salmonella typhimurium,* and this work typifies Demerec's research in the 1950s.

Demerec realized that transduction could be used to map the genes involved in tryptophan synthesis and began a project with his daughter

Zlata. Parallel studies were carried out on histidine mutants by Philip Hartman, who had come to the Department of Genetics as a Research Fellow of the U.S. Public Health Service. The first progress report appeared in the Carnegie Institution of Washington Yearbook in 1953 but by the time that Milislav and Zlata published the definitive account of the work in 1956, Zlata and Philip were married, and the paper is by M. Demerec and Z. Hartman.

The basic strategy in these experiments was to take two mutant strains with the same phenotype (e.g., a requirement for tryptophan), infect the "donor" bacteria with the transducing phage, isolate the phage and infect the "recipient" bacteria, plate out the infected bacteria on selective media, and count the recombinants. If the two strains carried mutations in different genes, complementation would occur and the recipient cells would grow in the selective medium.

A set of 10 *Salmonella* mutants that required tryptophan for growth were isolated, and pairwise transductions using all 10 mutants and wild-type cells were carried out. The power that bacteria and phage brought to genetics is shown by the numbers involved in these experiments. Transduction is a rare event occurring once in 100,000 infected cells, but Demerec and Hartman routinely added 1,000,000,000 phage to 200,000,000 bacteria.

Sydney Brenner spent the summer of 1954 with Demerec as a Fellow of the Carnegie Corporation. He had determined that the 10 mutants fell into four distinct biochemical groups, A through D, depending on which step was affected in the biosynthetic pathway to tryptophan. When Demerec and Hartman analyzed their results, they found that their genetic data were consistent with Brenner's biochemical findings.

They then turned to mapping the mutations. The data from these experiments enabled them to determine that the order of the four groups of *try* mutations along the chromosome, and from further experiments, they were able to determine the order of mutations *within* groups.

Philip Hartman's findings for mutations in the histidine locus were similar, and it was realized that the order of the loci on the chromosome corresponded with the order of the reactions in biochemical pathway. Demerec called this an "assembly line" arrangement and suggested that this close association came about as a consequence of evolutionary selection for the most efficient arrangement of genes. This was an early hint of the operon concept, developed in full and so successfully by Jacob and Monod, who cited Demerec's 1956 Cold Spring Harbor Symposium presentation in their classic paper.

Waclaw Szybalski Explores the Genetics of Antibiotic Resistance

In 1951, bacterial geneticist Waclaw Szybalski joined the Department of Genetics staff. He moved into Barbara McClintock's Animal House laboratory when McClintock, Hershey, and Kaufmann all moved into the new Carnegie laboratory.

Throughout his stay at the Biological Laboratory, Szybalski worked on bacterial resistance to antibiotics. Immediately after the announcement of the discovery of the new antibiotic isoniazid, he began testing the patterns of the development of resistance to isoniazid in three species of mycobacteria. A mutation increasing the resistance 1000-fold was found to occur in a single step, just as in the case of resistance to streptomycin. Szybalski further showed that by using two antibiotics together, the threat of mutation of a pathogen to resistance could be nullified, because the double mutation occurred in only one in a billion cells. There were practical implications of his work. There was no cross-resistance between isoniazid and streptomycin, suggesting that these agents could be used against tuberculosis. Szybalski found different patterns of penicillin resistance in *Staphylococcus*, indicating that physicians would need to take the patterns of cross-resistance into account for each variant agent of disease and for each different antibiotic.

Waclaw Szybalski at the 1951 Symposium.

In 1955, Szybalski left Cold Spring Harbor, moving first to Rutgers University and then to the University of Wisconsin—Madison. However, he maintained a close connection with Cold Spring Harbor over many years, attending every Phage meeting for more than 60 years. In 2004, Szybalski made a most generous gift to Cold Spring Harbor Laboratory that enabled the addition of the Szybalski Reading Room to the Library that formerly housed the laboratories of the Department of Genetics (Chapter 17).

McClintock Presents Her Work at the 1951 Symposium

McClintock had continued to pursue the strange behavior of the *Ds* and *Ac* transposable elements but reported her studies only in the Year Books of the Carnegie Institution of Washington. By 1950 she must have felt that she was on solid ground and in 1950 and 1951 presented her findings to a wider audience.

As a member of the National Academy of Sciences (elected in 1944; the third woman to join that august body), she submitted them to the Academy's *Proceedings.* She justified the absence of data, tables, and photographs: "The accumulated observations and data are so extensive… that no short account would give sufficient information to prepare the reader for an independent judgment of the nature of the phenomenon." McClintock added that "[m]anuscripts giving full accounts were in preparation," and the 1951 Symposium on *Genes and Mutations* provided the opportunity to give that full account.

Richard Goldschmidt opened the session by reviewing findings that he felt cast doubt on the standard view of the gene, arguing that in some way the whole chromosome comprised the unit of heredity. Edgar Altenburg responded vigorously: "I cannot agree with those who assert that the gene is a nebulous figure of the imagination." It probably did not help McClintock's cause that Goldschmidt enthusiastically cited her work in support of his position!

Barbara McClintock at the 1951 Symposium.

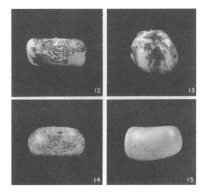

Examples of maize kernels from McClintock's 1951 Symposium paper.

The most interesting aspect of her presentation was what it revealed of McClintock's thinking about genes and genomes. She pointed out that she had deliberately avoided using the word "gene," and although she did not deny the existence of such units, she believed that the lack of knowledge of how these units operated in the chromosomes was such that "... no truly adequate concept of the gene can be developed until more has been discovered about the function of the various nuclear components." This was a view not likely to find favor in an audience largely made up of phage and bacterial geneticists whose careers were based on a belief that genes were defined physical entities.

McClintock emphasized the importance of *control* of gene expression: "The knowledge gained from the study of mutable loci focuses attention on the components in the nucleus that function to control the action of the genes in the course of development." It was for this reason that McClintock was later so enthusiastic about the work of Jacob and Monod (1961) that dealt with control of gene expression.

Accounts of the reception of her work at the Symposium are inconsistent. McClintock wrote that "[t]he response to it was puzzlement and, in some instances, hostility" and made similar comments in later interviews. For example, McClintock told Evelyn Fox Keller that "the Symposium in 1951 really knocked me." It appears that there was no discussion of her paper, although Norton Zinder recalled a separate meeting to discuss her work. However, Joshua Lederberg recalled a "lot of discussion," and Ernst Caspari, reviewing the Symposium in the pages of *Science* magazine, wrote that McClintock's talk generated "particular interest" in the debate over the nature of the gene. Whatever the truth of the matter, the fact remains that by and large McClintock reverted to publishing her work in the Carnegie Yearbooks.

Wallace's Studies on Radiation Prove Controversial

Drosophila had not been entirely abandoned at Cold Spring Harbor. Bruce Wallace joined the Laboratory in 1949 on a U.S. Atomic Energy Commission contract to examine the effects of ionizing radiation on the genetics of populations. The research was driven by three facts: Ionizing radiation had become an integral part of the industrial, military, and medical enterprises; radiation induces mutations; and nearly all such mutations are harmful. He hoped that his research would "… be of use in predicting the outcome of exposure of our own population to radiations…." James King joined Wallace soon after the project began, and they were aided by as many as six research assistants.

Wallace and King established seven populations of *Drosophila* exposed to various levels of radiation and found that widespread heterozygosity was strongly selected for in the irradiated populations. Many were "isoalleles," alleles that differ genetically but produce the same phenotype. Wallace concluded that mutations were not aberrations from an ideal "normal" genotype but variants whose value depended on the constellation of other alleles and environmental factors. A given allele might be beneficial in one organism or one population and detrimental in another.

Radiation genetics was an area of contention during these Cold War, post-Hiroshima years. Wallace's results were hotly disputed by H.J. Muller and other advocates of the "classical" theory of mutations, who thought that every radiation-induced mutation would be harmful. Wallace and King's government-sponsored research showing that not all radiation is harmful could raise eyebrows in antinuclear circles, and Muller feared their work would be seen as supporting the view that nuclear testing was not dangerous genetically. However, Dobzhansky, Wallace, and King agreed with Muller in one respect; all were opposed to nuclear testing.

Harold Abramson, LSD, and the CIA

Harold A. Abramson from the College of Physicians and Surgeons, Columbia University, and later Mount Sinai Hospital, was a regular of summer visitor to the Biological Laboratory. Abramson first became associated with the Biological Laboratory in 1933 when he became a member of the Scientific Advisory Committee. He worked during the summers on various biophysical topics, including electrophoresis of proteins, methods to get drugs through the skin, and allergies. During the war years, he led the Biological Laboratory's work on aerosols. After the war he retuned to his previous interests, but in 1953 his summer research at the Biological Laboratory took an entirely new direction.

Abramson had a program at Mt. Sinai Hospital studying the effects of LSD on human behavior. During the summer of 1953, he carried out research at the Biological Laboratory on the metabolic effects of LSD on guinea pig brain extracts. The next year he became a member of staff, and in 1956 his group was listed in the Biological Laboratory annual report as the "Psychobiology Section."

A goal of the work at the Biological Laboratory was to develop a bioassay for LSD. Abramson tested *Daphnia*, the water flea ("supplied by the local pet shop"), eight strains of *Drosophila*, the Mystery snail *Ampullibria cuprina*, and the Siamese fighting fish *Betta splendens*. The latter seemed to give the best results, reacting specifically to LSD25 with a very consistent "nose up–tail down" posture, which increased as the concentration of LSD increased in the water.

In the late 1950s, Abramson and collaborators at several different clinical institutions used LSD-induced states as a model for psychosis and schizophrenia, searching for drugs that would reduce anxiety levels. Abramson wrote that schizophrenic psychosis might well be treatable by repeated doses of psychomimetic drugs that in effect wear down the brain's ability to produce the psychotic state. Abramson investigated

Lysergic Acid Diethylamide (LSD 25):
II. Psychobiological Effects on the
Siamese Fighting Fish

Harold A. Abramson and Llewellyn T. Evans

*Biological Laboratory, Cold Spring Harbor,
Long Island, New York*

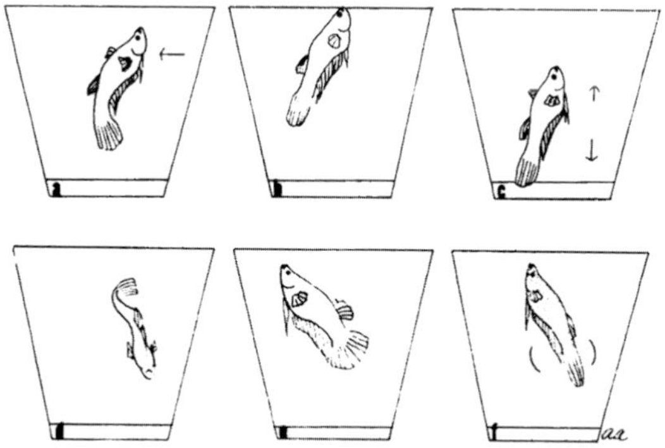

Fig. 1. The six diagrams illustrate the nine criterions or effects of LSD 25 upon the fighting fish, *Betta splendens* Such effects are essentially exaggerated postures or caricatures of the normal fish.

Title and diagram from Abramson's paper describing a bioassay for LSD using Siamese fighting fish.

psilocin from "magic mushrooms," and in his last report in 1961, described the effects of LSD on larger fish such as carp. This was the last report of the Psychobiology Section and was the end of Abramson's association with the Cold Spring Harbor.

In the 1970s, it was revealed there was a hidden side to Abramson's work. In his report for 1954–1955 and 1955–1956, Abramson acknowledged support from the Geschickter Fund for Medical Research and the Josiah Macy Jr. Foundation. Some 20 years later, during hearings held by Senator Ted Kennedy, the Fund and the Foundation were exposed as having been conduits for the CIA to fund research in academic institutions as part of its MKULTRA program. The program was to investigate

P&S prof conducted LSD tests

By ROBERT POLNER

Harold Abramson, assistant professor of physiology at Columbia from 1935 to 1958, was covertly furnished over $85,000 by the Central Intelligence Agency during the 1950s to experiment with LSD on human subjects. Abramson concealed his participation in the CIA's MKULTRA mind control program from the university and his research colleagues.

Now an allergist who practices in Manhattan, Abramson was a key figure in MKULTRA, an umbrella program under which many sensitive projects were funded. The CIA during the 1950s and 1960s intensely investigated the use of LSD as a possible truth serum and as a memory control and behavior modifying chemical.

Abramson's name became associated with MKULTRA when it was first disclosed in 1975, 22 years after the fact that he treated biochemist Frank Olson before his drug-induced suicide. Olson plunged through a closed tenth story hotel window in New York City after being given LSD unwittingly by CIA personnel in Maryland. Several days before his suicide, CIA operatives sent Olson to Abramson—who had CIA security clearance—after he exhibited a paranoid reaction to LSD.

A clear picture of Abramson's relations with the CIA and experimentation with LSD was

PHOTO COURTESY COLUMBIANA LIBRARY

Harold Abramson
. . . mind control

pieced together from scores of documents and interviews. Public records touching on his career include the text of the Senate Health and Science Subcommittee's 1975 and 1977 hearings on government financed drug experimentation, 40 articles and two books Abramson wrote about his work, and materials released by the CIA in response to a request under the Freedom of Information Act by John Marks, author of the just-published book **The Search for the Manchurian Candidate**. Abramson declined to be interviewed during the preparation of this article, and the CIA has refused to release its files on MKULTRA activities at Columbia.

President McGill said he has never heard of Harold Abramson. Abramson's name wasnot mentioned in materials the CIA released to McGill in 1977.

"In the late 50s and early 60s," McGill said in an interview in March, "there was some contact between the CIA and the medical school, and I base that on reports given me
See ABRAMSON, 11

Article by Robert Polner on the front page of the Columbia Daily Spectator *from May 1979, reporting on the revelations concerning Abramson and the CIA's MKULTRA project.*

how mind-controlling drugs, including LSD, could be used in interrogations, and to learn how such drugs could be resisted.

Clouds on the Horizon

The 1950s had begun with an air of justified optimism at Cold Spring Harbor, an optimism that was fulfilled in those years. Scientifically and financially Demerec's institutions were at their peak. In the immediate postwar years, the staff of the Biological Laboratory numbered 14; by the mid-1950s it was 40. Its budget had remained fairly stable, overall generating a surplus between 1950 and 1955. The Carnegie Institution of

Washington had continued to provide stable funding for the Department of Genetics.

However, there were clouds on the horizon, clouds that became darker as the decade progressed. In 1960, Demerec was going to be 65, the mandatory age for Carnegie Institution of Washington employees to retire, and it seemed that the latter part of the decade, leading to his retirement, was going to be challenging.

HERSHEY AND CHASE SHOW THAT THE HEREDITARY MATERIAL OF BACTERIOPHAGE IS (PROBABLY) DNA

Summary

Al Hershey and Martha Chase's "Waring Blendor" experiment was a major step in persuading doubters that DNA, and not "genetic protein," is the hereditary material.

The Discovery/The Research

Alfred Hershey began his research career in phage research in the 1940s and came to Cold Spring Harbor in 1950. By then he was closely associated with Max Delbrück and Salvador Luria, the three forming the core of the phage group. Hershey was interested in phage replication and decided to exploit new electron microscopy findings that a bacteriophage attached to a bacterial cell and seemed to "inject" something into the bacterium it was infecting.

With Martha Chase as his research assistant, Al Hershey began the experiment by labeling phage T2 protein with ^{35}S and the DNA ^{32}P. These phage were added to a suspension of bacterial cells, which were then spun for a few minutes in a Waring Blendor to strip the phage from the bacteria cell wall. The suspension centrifuged briefly to pellet the bacteria. They found that most of the ^{32}P-labeled DNA was associated with the bacteria, whereas most of the ^{35}S was in the supernatant. However, of the 20% of ^{35}S that could not be removed from the cells, little or none could be found in the newly replicated phage particles. On the other hand, much of the ^{32}P that entered in the infected cells could be recovered in the new phage.

They concluded that they had shown that a "physical separation of the phage T2 into genetic and non-genetic parts is possible." The interpretation was that it was the phage DNA that was the genetic material of the virus, although Hershey and Chase were more cautious: "The DNA has some function. Further chemical inferences should not be drawn from the experiments presented."

Significance

In 1944, Oswald Avery and his team at Rockefeller University had carried out experiments on pneumococcus that provided very strong evidence that DNA was the hereditary material. However, many scientists still believed in the "genetic protein." DNA, they said, was only a scaffold for the genetic protein. The Hershey–Chase experiment helped convince these skeptics that hereditary material was DNA and not protein, even though it was less rigorous than those of Avery et al. It was more persuasive, perhaps, because it was carried out at a time when the tide was turning in favor of DNA and was in the mainstream of the developing field of molecular genetics.

Martha Chase and Alfred D. Hershey, 1953.

Continued

The Scientists

Alfred Hershey: Hershey was born on December 4, 1908 in Owosso, Michigan. He earned his B.S. degree and Ph.D. at Michigan State University before moving to Washington University, St. Louis. Hershey came to the Department of Genetics at Cold Spring Harbor in 1950, and became Director of the Genetics Research Unit in 1962. He shared the 1969 Nobel Prize in Physiology or Medicine with Max Delbrück and Salvador Luria. Hershey died on May 22, 1997, at his Laurel Hollow, New York, home.

Martha Chase: Chase was born on November 30, 1927 in Cleveland, Ohio. She graduated in 1950 from The College of Wooster in Ohio and was appointed research assistant to Al Hershey the same year. She resigned her Cold Spring Harbor position in 1953 and then worked at the Oak Ridge National Laboratory and the University of Rochester. In 1959 she began her Ph.D. at the University of Southern California, which was awarded in 1964. After working at the University of Southern California, she returned to her home in Cleveland. She died August 8, 2003, in Lorain, Ohio.

— *J.C.*

WITKIN'S ULTRAVIOLET-INDUCED BACTERIAL MUTATIONS OPEN THE DOOR TO EXCISION REPAIR AND THE SOS RESPONSE

Summary

Witkin studied biological responses to DNA damage, specifically using ultraviolet (UV) light–induced mutations in bacteria and deduced the error-prone repair mechanisms responsible for DNA repair.

The Research/The Discovery

When Evelyn Witkin was a graduate student spending the summer of 1944 at Cold Spring Harbor, Milislav Demerec gave Witkin a UV General Electric germicidal lamp and said, "Go, induce mutations." In her first experiment, Witkin identified the B/r strain of *E. coli*, which was more resistant to radiation than the parental B strain, the first time a mutation conferring increased radiation resistance had been identified.

Witkin returned to Cold Spring Harbor the next year to complete her thesis, which focused on understanding the genetics of UV radiation resistance. Once on staff at the Department of Genetics at Cold Spring Harbor, Witkin found that mutation frequency declined if protein synthesis was delayed or somehow inhibited. She found that adding amino acids (e.g., tryptophan) to the medium suggested the importance of protein synthesis. Adding chloramphenicol—an inhibitor of protein synthesis—reduced the yield of mutants. It appeared that some genetic repair mechanism was involved—what Witkin initially referred to as process "X." One possibility was that X is the process of "repair of genetic damage" and that protein synthesis favored the repair of this damage. Witkin called this effect mutation frequency decline (MFD).

Witkin believed that the UV-damaged DNA might produce a coordinated cellular response. She speculated that there was a "dark repair" mechanism that was complementary to photorepair by visible light (discovered by Oscar Kelner [Chapter 10]) and isolated a mutant that was defective in this process. In 1964, Richard Setlow and W.L. Carrier showed Witkin's dark repair to be DNA excision repair. In 1966, Witkin identified her process X as excision repair of pyrimidine dimers in genes that code two tRNA molecules.

In the early 1970s when she was at Rutgers, Witkin (and others in the field) received a letter from Miroslav Radman at Harvard suggesting that the Weigle phenomenon and UV mutagenesis of bacteria occur by what he called SOS replication. He proposed the idea of induction of SOS mutagenesis by SOS replication and suggested approaches to identify SOS activation. Witkin and Radman added putative UV-inducible functions and determined that their regulation depended on *recA* and *lexA*, thus showing that SOS mutagenesis is an inducible component of the SOS response.

Significance

Witkin's ideas were formulated before DNA was known as the genetic material or that UV irradiation damaged DNA directly. Her discovery that UV killing was related to control of cell division in the *E. coli* B strain is now known as "checkpoint" control, a point in the cell cycle during which cell division is inhibited until genome damage is repaired.

Witkin's conclusion that UV DNA damage activated a coordinated cellular response is now known as the SOS response. Subsequent research in the 1980s identified the nature of the SOS response of DNA to mutagenic activity. The response ultimately produces the production of error-prone DNA polymerases that fill in the lesions so that replication can continue; but they also introduce sequence changes, or mutations.

Continued

William Hayes, Evelyn Witkin, and Norton Zinder at the 1953 Symposium.

The Scientist

Evelyn Maisel Witkin was born on March 9, 1921, in New York City. She earned her B.A. degree in 1941 at New York University and went on to do graduate studies at Columbia with Theodosius Dozhansky. Instead of chosing Dobzhansky's flies to study, she chose *E. coli*.

She received her Ph.D. in 1947. From 1947 to 1955 she was a Staff Scientist at the Department of Genetics in Cold Spring Harbor, and of her time there she said "...it was heaven." She was at State University of New York Downstate Medical Center in Brooklyn from 1955 to 1971, and then she moved to Rutgers University as Professor. She was named Barbara McClintock Professor of Genetics in 1979 and in 1983 went to the Selman Waksman Institute, Rutgers University, retiring in 1991. She was elected to the National Academy of Sciences in 1977. Her many awards include the 1982 American Women of Science Award for Outstanding Research, 2000 Thomas Hunt Morgan Medal, and 2002 National Medal of Science.

— *J.C.*

Troubled Times

The story of the Cold Spring Harbor laboratories in the late 1950s is less concerned with research than with survival as the momentum of the previous years began to wane. Beginning with Evelyn Witkin in 1955, over the next several years there were resignations that weakened the research programs of both the Biological Laboratory and the Department of Genetics. There was change elsewhere, too. In 1955, Vannevar Bush retired at the Carnegie's mandatory retirement age of 65. Bush had been a great friend of both Cold Spring Harbor laboratories, and although his replacement, Caryl Haskins, served admirably, Bush's retirement marked the end of a close relationship between the Carnegie Institution of Washington and Cold Spring Harbor. With hindsight, this change at the Carnegie appears more ominous than it must have at the time. The Biological Laboratory seemed on a more secure footing than in previous years, although the annual reports almost always contained a plea for endowment or warnings about the poor state of the physical plant of the Biological Laboratory.

Vannevar Bush at MIT in an undated photo.

Symposia, Meetings, and Courses

One bright spot for the Biological Laboratory was the education program. The Symposia continued to be a highlight of the Cold Spring Harbor year, although several were on topics rather far from the microbial genetics which was at the forefront of biological research. The Phage (now Bacterial Viruses) and Bacterial Genetics courses continued, and a new course on tissue culture was added in 1958. The Nature Course for "young people," begun in 1941, was still very popular—enrollment in 1959 reached 195. That same year, there was a new addition to the education program when the National Science Foundation (NSF) made a grant for an Undergraduate Research Program. There were 10 students in the first class, including future Nobel Prize winner David Baltimore. It was a great success, and as Arthur Chovnick wrote in the Biological Laboratory annual report,

> It was quite clear from the enthusiasm of the staff and students, and from the remarkable intellectual growth exhibited by the students during the short duration of the program, that it would be well to continue this activity in the future.

And continue into the future it has, despite some inevitable difficulties with funding. NSF withdrew funding in 1973 and the program was funded through a variety of sources until 1987 when the NSF funding was renewed. The program was put on a firm footing with a $1,000,000 endowment by the Burroughs Wellcome Foundation, and with gifts from other sources, the Undergraduate Research Program Fund totals more than $3 million. The number of undergraduates coming to Cold Spring Harbor each year now numbers about 25.

Undergraduate Alan Rein and Arthur Chovnick, Biological Laboratory Director, examining Drosophila *in 1961, the second year of the Undergraduate Research Program.*

Changing Staff

Evelyn Witkin resigned in August 1955, for a post at SUNY Downstate Medical Center in Brooklyn. She had come as Dobzhansky's graduate student to Cold Spring Harbor in 1944 and became a member of staff in 1955. Witkin had helped organize and teach a popular summer course in bacterial genetics, begun in 1950. She was also a founding editor of the *Microbial Genetics Bulletin,* modeled on the *Drosophila Information Service.* Vernon Bryson left the Biological Laboratory in the fall, taking a leave of absence to work at the National Science Foundation in Washington. He did not return to Cold Spring Harbor, instead resigning his post at the Biological Laboratory to become a professor at Rutgers University and assistant director of its Institute of Microbiology. In 1958, two more long-term scientists left. Bruce Wallace resigned from the Biological Laboratory to take an academic post at Cornell University, and James King moved his research program from the Department of Genetics to Columbia University.

Demerec hired several staff members to replace Witkin, Bryson, Wallace, and King, most working in bacterial and phage genetics and concentrating on "hot" areas such as radiation and physiological genetics. Ellis Englesberg, for example, joined the Laboratory staff in 1955 to pursue nutritional mutations in bacteria. Similar research was carried out by new staff members Paul Margolin and Abraham Schalet, who used *Salmonella* and mutagenesis to study the biochemical genetics of the biosynthesis of leucine and tryptophan, respectively. Palmer D. Skaar took Bryson's place as head of the genetics of bacteria section, studying the nature of genetic recombination by means of conjugation between different strains of *Escherichia coli*.

In 1957, Hermann Moser, who for 2 years had been a fellow of the Carnegie Institution in the Department of Genetics, moved from the Department of Genetics to the Biological Laboratory. Moser began working on mammalian cells in culture, a new area of research at Cold Spring Harbor. It was a forward-looking choice by Demerec as cell culture was just becoming a useful tool with the development of standardized techniques. Moser established a tissue culture facility and in 1958 began a course on the Quantitative Study of Human Cells in Tissue Cultures. A notable student in the first course was Purnell Choppin, who later became President of the Howard Hughes Medical Institute.

These new researchers were all competent scientists but they made little impact on the Biological Laboratory or the Department of Genetics. An exception was Arthur Chovnick, who played a role in the crisis that followed Demerec's retirement.

Research in the Department of Genetics

Demerec was by this time an elder statesman of science. He participated in a National Academy of Sciences committee, funded by the Rockefeller Foundation, the Atomic Energy Commission (AEC), and the Department of Defense, to examine and report on the biological effects of atomic radiation. It had some of the leading geneticists, including Kaufmann, Demerec's colleague in the Department of Genetics.

Demerec's participation on the AEC committee influenced his research program. It had never been established whether the linear relation between radiation dose and mutation rate extended to very low doses. If it did, then exposure even to the infrequent doses of X rays used by physicians, dentists, and shoe salesmen would produce a proportional increase in the number of harmful mutations. Demerec undertook to determine this by irradiating bacteria at varying intensities of X rays and counting the number of mutants. He found that, at least to doses down to 8.5 r, the linear dose/response relationship held true. Demerec also

---

MEMBERSHIP OF THE COMMITTEE ON

GENETIC EFFECTS OF ATOMIC RADIATION

GEORGE W. BEADLE, California Institute of Technology, *Chairman*
H. BENTLEY GLASS, Johns Hopkins University, *Rapporteur*
JAMES F. CROW, University of Wisconsin
M. DEMEREC, Carnegie Institution of Washington, Cold Spring Harbor, L.I., N.Y.
THEODOSIUS DOBZHANSKY, Columbia University
G. FAILLA, Columbia University
ALEXANDER HOLLAENDER, Oak Ridge National Laboratory
BERWIND P. KAUFMANN, Carnegie Institution of Washington, Cold Spring Harbor, L.I., N.Y.
H. J. MULLER, Indiana University
JAMES V. NEEL, University of Michigan
W. L. RUSSELL, Oak Ridge National Laboratory
A. H. STURTEVANT, California Institute of Technology
SHIELDS WARREN, New England Deaconess Hospital, Boston
SEWALL WRIGHT, University of Wisconsin

Membership of the National Academy of Sciences committee on The Biological Effects of Atomic Radiation.

showed that different genes have different sensitivities to X rays. This work occupied Demerec up to his retirement, his last research report appearing in the annual report for 1957–1958.

Hershey's group continued studying the respective roles of DNA and protein in phage T2, describing the fine points of the blender experiment. It was good, solid science; not revolutionary although interesting to specialists. Hershey was honored in 1958 with the one major scientific accolade for Cold Spring Harbor during these years. Together with Delbrück and Luria, he was elected to the National Academy of Sciences.

Barbara McClintock pressed onward in her studies of transposable elements in maize, and, in 1955, she identified a new transposon, which she called *Spm*, for suppressor-mutator. Her conviction that these elements controlled differentiation and development grew stronger with each growing season. In 1956, she made another attempt to persuade her peers of the role of transposable elements, at the 1956 Symposium on *Genetic Mechanisms*. The reception in 1956 was perhaps even cooler than in 1951. McClintock was devastated. It was the last public presentation she would make of the primary data on transposable elements.

Perhaps frustrated by the limited impact of her transposon studies on the world of genetics, McClintock made two trips in 1957 and 1959 to trace the spread of various races of maize through South America. Could physical features on the chromosomes be used as markers to trace the evolution of various races of maize? She focused her attention on the chromosome "knobs" she had used for many years as markers. Using these she found that the races of maize in cultivation in the highland regions of Ecuador, Bolivia, Chile, and Venezuela were remarkably similar. McClintock produced elegant and detailed maps of the various races of maize, which could be used to trace the origins and migrations of maize and the effects of hybridization. The project must have had considerable appeal for McClintock, emphasizing observation, patterns, and natural history, and providing what must have been a welcome break

Al Hershey in his laboratory.

Barbara McClintock with Harriet Creighton at the 1956 Symposium, Genetic Mechanisms: Structure and Function.

Sent to Blumenschein, July 15, 1963

Six plants examined in each race:

Name of Race	Collection Number	Origin of Seed for Plants Examined	Number of plants with distal knob in chromosome 10.		Number of plants with Abnormal 10 (Each plant heterozygous for abnormal chromosome 10)
			Homozygous	heterozygous	
Reventador	Nay. 36	Ji-51	1	1	0
Conico Norteño	Gto. 102	Original	1	0	0
Maíz Dulce	Gto. 100	Original	0	3	0
Celaya	Celaya II	Ji-52	0	2	1
Palomero Toluqueño	Mex.210	Original	2	2	0
Chalqueño	Mex.208	Original	0	1	0
Bolita	Oax. 180	Original	0	2	1*.
Imbricado	Guat. 720	Original	0	1	0
Jala	Nay. 72	Original	0	0	5
Arrocillo Amarillo	Pue. 260	Original	0	0	2
Zapalote Grande	Chis. 236	Original	0	0	2
Salvadoreño de C.R.	C.R. 30	Original	0	0	4

*This plant had the distal knob in chromosome 10, in the homologue.

McClintock's data on the number of plants of different Mexican races of maize with a distal knob in chromosome 10.

Berwind Kaufmann demonstrates the new electron microscope to David Baltimore, a student in the first class of the Undergraduate Research Program, 1959.

from trying to convince colleagues of her increasingly complex view of the genome.

Hermann Moser's tissue culture was not the only new technique to appear at Cold Spring Harbor in this period. The National Science Foundation provided funds for an electron microscope, and during 1958 the microscope passed its initial performance tests and came into full use. Although the grant was to the Biological Laboratory, Berwind Kaufmann in the Department of Genetics helped make the instrument ready for general use and used it extensively in his cytogenetic studies. Kaufmann, in fact, became the resident electron microscopy expert and in the coming years assisted many Cold Spring Harbor scientists with electron microscopy projects.

Demerec Retires

Although only the Carnegie Institution imposed mandatory retirement, Demerec had decided that if he had to retire from the Department of

Genetics, he would also retire as director of the Biological Laboratory. Unlike Charles Davenport, who retired from the Laboratory in 1924 and the Carnegie in 1934, Demerec, who had worked to promote collaboration between his charges, would not cleave the two laboratories by retaining the directorship of one. The Carnegie agreed to provide Demerec with a 3-year grant beyond his retirement of the directorship, during which he and his group could continue experiments.

As a first step, a new assistant director was needed to replace Bruce Wallace. The LIBA Board of Directors settled on Arthur Chovnick from the University of Connecticut. Chovnick had attracted attention for his work on the fine structure of complex, or compound, genes in *Drosophila*. Chovnick arrived in January 1959 and as assistant director was successor-apparent to Demerec.

Arthur Chovnick, 1960.

The forthcoming change in leadership must have heightened the Board of Trustees's awareness of the poor financial situation of the Biological Laboratory. It had been causing concerns as early as the mid 1950s when the departure of Vernon Bryson in 1956 caught the attention of Amyas Ames, LIBA's president. In his remarks in the 1955–1956 Annual Report, Ames wrote that, "He [Bryson] did not seek a larger salary. He sought security to pursue his researches without interruption, security for himself and his family." Ames pled for the raising of half a million dollars in research capital to provide the Laboratory's leading scientists with a stable source of funding.

In 1957, Walter Page, Ames's successor, formed a Special Policy Committee to "review the activities of the Association, make recommendations as to its future policy, and suggest means for putting such plans into operation." The Committee met with Caryl Haskins, Bush's replacement as President of the Carnegie Institution of Washington, to discuss the relationship between the two institutions. In 1958, Demerec presented a memorandum to the Board on the "Present Status and Plans for Future Development" of the Biological Laboratory. In 1959, LIBA's

Amyas Ames.

Board of Directors persuaded a group of distinguished scientists to be the Science Policy Review Committee. The group consisted of Edward Tatum, chairman, Salvador Luria, Ernst Caspari, James D. Watson, Edgar Zwilling, and William D. McElroy. They were to aid "the Board of Directors in determining a wide and constructive policy to insure a vigorous, outstanding and broadening program in the future."

In August 1959, the Committee issued a statement bravely titled, "Creative Adaptability." The report reviewed the past two decades of the Laboratory's history under Demerec's direction and approved the general direction of its policy as he had formulated it, especially the close coordination of its programs and aims with those of the Carnegie Department of Genetics. The committee emphasized the unique features of the program of the Laboratory: its symposia, its teaching programs, its summer research, and its successful community projects. It stressed also the desirability of continuing the Laboratory's policy of maintaining flexibility around its core of microbial genetics. "By flexibility of program," the committee said, "we do not mean 'jumping on every bandwagon,' but a sort of creative adaptability depending on the foresight and leadership of the entire staff, especially the director." Demerec's successor would bear a heavy responsibility.

The conclusion of the statement by the Science Policy Review Committee, August 1959.

The committee urges, therefore, that the past record of joint accomplishments be taken into account in planning for the future of the Cold Spring Harbor laboratories, and feels it to be of vital importance that problems related to the immediate and future development of the laboratories be considered jointly by representatives of both organizations concerned.

E. L. Tatum, *Chairman*

E. Caspari, H. B. Glass, S. E. Luria, J. D. Watson, E. Zwilling. Absent: W. D. McElroy.

The final version of Demerec's 1958 report is dated July 1959 and stresses the points made by others. The Biological Laboratory had an international reputation for its research, for the Symposia, and for its courses. Demerec's description of the importance of the meetings rings as true today as it did in 1959:

> The atmosphere of the Laboratory is well adapted to such informal contacts, which are the unique and perhaps most valuable function of our Symposia. Many friendships have been started at the Symposia, and many scientific controversies settled; ill feelings have been reconciled, and numerous cooperative projects initiated. These gatherings contribute very substantially to the integration of biological research on a worldwide basis.

However, the Biological Laboratory's infrastructure was in a deplorable state. Of the five laboratory buildings, only two could be used throughout the year. James Laboratory, built 30 years earlier, had never had its second story added and the roof leaked. The Laboratory was using the Department of Genetics's former animal house, and although this had been built in 1916, it had better facilities than any building belonging to the Biological Laboratory.

As for the future, the LIBA Board had decided that Biological Laboratory should have its own director, paid by and responsible to LIBA:

> Because ... it is becoming evident that a good deal of time and thought are required for its administration—much more, in fact, than can justifiably be expected of a director whose salary comes from another institution and whose primary obligations are to that institution and to his own research.

On June 30, 1960, Demerec retired as director of both the LIBA Laboratory and the Carnegie's Department of Genetics.

The Long-Islander.

Huntington, L. I. N. Y   Thursday, June 30, 1960

DR. DEMEREC LEAVES

Demerec's retirement is announced in the local newspaper, The Long-Islander.

Post Demerec

The question of Demerec's successor at both the Department of Genetics and the Biological Laboratory raised concerns outside Cold Spring Harbor. In March 1959, Luria wrote to a group of scientists long associated with Cold Spring Harbor, including Delbrück, Dobzhansky, Glass, Lederberg, Mayr, Muller, and Rhoades, expressing his fears. He was worried that "... neither Demerec, nor his present staff, nor any representative group of geneticists has been asked to define goals and criteria for the selection of a Director for Carnegie" and that Haskins alone would decide the appointment. Luria asked whether steps should be taken by an outside group and, if so, what those steps should be.

A letter from Luria suggesting that action needed to be taken over the appointment of Demerec's successor.

MASSACHUSETTS INSTITUTE OF TECHNOLOGY
DEPARTMENT OF BIOLOGY
CAMBRIDGE 39, MASSACHUSETTS

March 6, 1959

TO: Dr. G. W. Beadle   Dr. J. Lederberg
Dr. E. Caspari   Dr. H. J. Muller
Dr. M. Delbruck   Dr. M. M. Rhoades
Dr. Th. Dobzhansky   Dr. H. Roman
Dr. L. C. Dunn   Dr. T. M. Sonneborn
Dr. N. Giles   Dr. C. Stern
Dr. B. Glass

Dear Friends:

The purpose of this letter is to raise some questions about the future of Cold Spring Harbor as a center of genetics. It is likely that a decision will soon be made as to Demerec's successor at Carnegie; this will, of course, affect the Biological Laboratory as well.

However awkward it is to butt into the affairs of another Institution, I wonder whether we, who have all been associated more or less closely at various times with Cold Spring Harbor's activities, are not entitled to express to Haskins the concern of geneticists in this matter. One possible suggestion may be the appointment of an ad hoc advisory committee.

The response was almost unanimous—nothing should be done. In the event, Berwind Kaufmann was named Acting Director of the Carnegie Department of Genetics, becoming director in 1961. This could only be a caretaker measure as Kaufmann would reach the Carnegie mandatory retirement age in 1962.

Arthur Chovnick became director of the Biological Laboratory; but although the transition from Demerec to Kaufmann was apparently smooth, that from Demerec to Chovnick was not. Demerec appears to have persuaded the Board of Trustees that as retired director he should have life tenure while Chovnick should be on an annual contract. Chovnick could not accept this—his position as director would be compromised. Instead the Trustees gave Demerec an annual contract and Chovnick a renewable two-year contract.

Difficulties continued over an addition to James Laboratory. In a letter to Walter Page on December 8, 1959, Demerec wrote that many had appealed to him to stay on at the Laboratory and he would do so on two conditions: first, that he would have no administrative duties and, second, that he would have laboratory space with the Biological Laboratory, not with the Department of Genetics. Demerec pointed out that James Laboratory had originally been designed as a two-story structure. If a second floor could be built with a laboratory for him, he would stay on. He said that he met with an architect and with Chovnick (then his Assistant Director), and everyone seemed to be comfortable with the idea of Demerec's occupying the proposed second floor of James. However, it seems that by the time Chovnick became director he had recognized that Demerec's continuing presence at the Laboratory would be disruptive. Chovnick claimed the second floor of James for his laboratory effectively forcing Demerec out. At this point, Demerec sought a post outside Cold Spring Harbor and received an offer from Brookhaven National Laboratory.

The second floor of James laboratory under construction.

He worked at Brookhaven for 6 years until in 1966 he was offered a research professorship at the C.W. Post campus of Long Island University. Demerec was greatly looking forward to moving to a new laboratory, and by the spring of 1966 it was nearly complete. But he never moved into it. On April 12, 1966, Demerec died of a heart attack. The Laboratory memorialized him in the Annual Report of 1966:

> Numerous legends tell of his flair for achieving great objectives with the simplest of means. Less well known are the difficulties he must have had to overcome... . The Laboratory itself, as it is known today to thousands throughout the world, remains the proper monument to his unfailing endeavor.

Chovnick Becomes Director of the Biological Laboratory

Chovnick assumed the directorship of a laboratory in a precarious financial position. In light of the Laboratory's financial straits, a $470,000 proposal to the NIH for a new 60-person, 25,000 square foot laboratory was shelved—the Laboratory simply did not have the capital to equip, staff, and maintain such a large new building. Instead, there was to be a $100,000, 4,000 square foot addition to James laboratory. The NIH offered a matching grant of $50,000, but even so the James addition was a financial disaster for the severely taxed Laboratory. In 1959–1960 it posted a surplus of nearly $20,000, but it had committed $70,000 to renovations. To complete the construction, the Laboratory raided and seriously depleted its operating capital.

During the summer of 1960, Chovnick wrestled with how to manage his new charge and particularly how to continue collaborations with the Carnegie Department. Demerec had written in his 1959 memorandum of the "clearly evident feeling of animosity" between the staff of the two Cold Spring Harbor institutions that had existed in 1941, prior to his assuming directorship of both. Perhaps Chovnick worried that such animosity might reemerge with the institutions no longer led by the same director. On June 7, 1960, Chovnick wrote to Kaufmann stating these opinions and his belief that Demerec "should have no role in determining the policy between our organizations after July 1, 1960." Demerec had been such a forceful presence for 20 years that Chovnick feared he would try to interfere in Chovnick's running of the Laboratory.

Despite this inauspicious beginning, Chovnick was able to make significant improvements, including introducing a pension scheme for the staff and improving food in Blackford by bringing in a new cook. Chovnick changed the company printing the Symposia volumes, resulting in a considerable improvement in the quality of the printing, and by more active advertising of the volumes, greatly increased sales. Most importantly for the following generations of resident and visiting scientists, it was Chovnick who first obtained a liquor license for a bar in the basement of Blackford Hall, open from 4 p.m. to midnight.

Chovnick's main contribution was to the community of Cold Spring Harbor. In 1958, the Town of Oyster Bay, of which Cold Spring Harbor is a part, considered dredging the harbor as part of a plan to develop the Cold Spring Harbor area. This was to include a public marina for the sand spit on which so many scientists relaxed in the summer—sunning, swimming, and talking science. The plan alarmed the scientists. Chovnick and Demerec campaigned locally, writing letters and making presentations to local chapters of environmental groups such as the Nature Conservancy. Letters of support poured in from prominent scientists worldwide, heaping shame on the Town of Oyster Bay for selling out such an important scientific institution for the sake of a buck. Many of these pleas invoked the Laboratory's long tradition of marine ecology research, although, as the Town officials pointed out, it had been many years since marine biology or ecology research was done at Cold Spring Harbor. But this did suggest a strategy to thwart the Town, and, in 1959, Chovnick proposed that the Laboratory take up a new program of marine physiology. On August 11, 1960, the Cold Spring Harbor Marine Biological Reserve was established, protecting the harbor from development. This was not, however, a permanent solution and the issue rose again 12 years later (Chapter 16).

Other than this environmental victory, there was little to celebrate in 1960. The staff was depleted: Only Chovnick, Margolin, Moser, Schalet, and Abramson remained, and only Abramson had any research to report. Years of neglected maintenance had taken a severe toll on the buildings. A February 1960 memo to the trustees reported that Blackford Hall and Jones Laboratory required "some modifications"; James needed "extensive changes"; Wawepex was unusable, "effectively condemned"; condemnation of Davenport Laboratory and Williams House was imminent; and Hooper House was salvageable but only with extensive changes.

In 1960, John Cairns, a British scientist working at Australian National University, spent a sabbatical year at Cold Spring Harbor with Al Hershey. Later, he described eloquently his first impressions of the Laboratory grounds:

[T]he first face Cold Spring Harbor presented to the visitor was one of genteel decay. All those dilapidated wood-frame buildings, peeling and rotting in the heavy heat of a Long Island summer; the occasional musty brick library or laboratory standing on its own, as testimony to some bygone burst of optimism and affluence; fields of unmown grass; an undredged harbor, filling the air with the smell of fermenting mud—in short, a landscape wrapped up in its past and apparently no longer being touched by living hands. Newly arrived, you noticed that the people you met all seemed equally alienated, as though you had stumbled into the middle of a Fellini movie.

A Crisis and the Founding of the Cold Spring Harbor Laboratory of Quantitative Biology

Two cataclysmic events brought an end to the two research institutions that had been at Cold Spring Harbor for so long and led to the creation of a new entity.

In 1962, Berwind Kaufmann retired, forcing the issue of whether there should be a Carnegie Department of Genetics at all. The research in the department was no longer at the cutting edge of genetics, which, by the early 1960s, was focused on molecular studies of the "central dogma" (i.e., how information in DNA was used to make proteins). Other departments of the Carnegie were more attuned to the times, among them, remarkably, the Department of Terrestrial Magnetism. It had had a section on biophysics since late 1940s, initially using radioactive isotopes to trace biochemical reactions. By the late 1950s, Dick Roberts and Roy Britten were working on protein and nucleic acid synthesis as well as the structure of the ribosome, research much more in line with the hot topics in molecular biology than the research at Cold Spring Harbor. The Departments of Plant Biology and of Embryology were also deeply involved in molecular biology and genetics, and perhaps it seemed less important to maintain a separate Department of Genetics at Cold Spring Harbor.

```
┌─────────────────────────────────────────────────┐
│ GENETICS RESEARCH UNIT                           │
│                          Cold Spring Harbor      │
│                     Long Island, New York 11724  │
│                                                  │
│           Alfred D. Hershey, Director            │
│           Elizabeth Burgi                        │
│           Barbara McClintock                     │
│           Margaret R. McDonald                   │
│                                                  │
└─────────────────────────────────────────────────┘
```

The members of the Genetics Unit as listed for the first time in the Carnegie Institution's Yearbook 62 for 1962–1963.

The Carnegie Institution's decision was compassionate but firm. The Department of Genetics would be closed but a "Genetics Research Unit" would be established that would be funded as long as McClintock and Hershey remained active in research.

The Biological Laboratory was facing a crisis of its own. In 1962 Chovnick resigned his post at the Laboratory and returned to the University of Connecticut. His research was going well and he was not yet ready to give up research for administration. Moreover, the Board of Trustees was not giving him the support he needed. The financial success of the Symposia volumes had led him to propose expanding scientific book publishing, arguing that selling books could become as profitable to the Biological Laboratory as selling mice was for the Jackson Laboratory in Bar Harbor. Chovnick also developed a plan for additional laboratories, but the Trustees were not enthusiastic about either initiative. Edwin Umbarger, who had come to the Laboratory in 1960, was appointed interim director while a permanent director was sought.

Whether there would be an institute to direct was another matter. The Laboratory had been in trouble before, but never trouble this bad. It was reported at the March 1961 meeting of the LIBA Board of Directors that capital expenses exceeded income by $109,000 and that LIBA's unrestricted funds were reduced to $25,000. Destitute both financially and intellectually, and with the loss of the Carnegie Department, it seemed to some that the only option was to let it go bankrupt, which might enable a bailout. Eleventh-hour letters of support solicited by Francis Ryan came from scientists around the world. A generation of scientists had "grown up" scientifically at Cold Spring Harbor, and when they heard of its troubles they expressed sadness and outrage that the institution might fail. A "White Paper" prepared by the Laboratory's Scientific Advisory Committee noted that:

> This concern is not only sentimental. It comes from a realization that the geographic location of CSH, in a sheltered setting at the gateway of the US, provides unique facilities for international contacts among biologists. CSH is not only a symbolic but also a physical center of gravity, which enhances the leading role of our country in biological science.

The Carnegie Institution was willing to turn over its land and buildings to the Laboratory, but there had to be a Biological Laboratory to receive them. The Scientific Advisory Committee sought alternatives. One proposal was for the Laboratory to be operated by Associated Universities, a collective of regional universities that managed Brookhaven National Laboratory. There was discussion of a university taking over but these plans foundered as had others in the past for fear that external control would diminish the Laboratory's distinctive character.

Instead, the Scientific Advisory Committee devised a plan for a new corporation, the Cold Spring Harbor Laboratory of Quantitative Biology (CSHL), which would be funded by a group of universities and scientific institutions. The participating institutions, which included The Rockefeller University, Duke University, Albert Einstein College of Medicine, Princeton, Brooklyn College, NYU, Sloan-Kettering Institute for Cancer Research, and the Public Health Research Institute, would each name one representative to the CSHL's Board of Trustees. The Long Island Biological Association (LIBA) would become one of the participating institutions, its primary role in CSHL reduced to fundraising and other support, and have two trustees. The Wawapex Society made over its land and buildings to the new laboratory, and it, too, would have a seat on the Board of Trustees.

John Cairns Becomes the First Director of the Cold Spring Harbor Laboratory

For the first time since Reginald Harris, the Laboratory sought its new director from outside the scientific staff—it may well be that no one currently on the staff wanted the job—and found their man in John Cairns.

Born in England and trained at Oxford University, Cairns had received an M.D. in 1946. After his postdoctoral training, he joined the faculty of the Australian National University and remained there for 8 years. In 1960, Cairns brought his family halfway around the world to spend a year in Hershey's laboratory. At that time, although it was clear that the phage T2 chromosome contained a single piece of DNA, it was not known whether this was a single continuous double helix. During his year with Hershey, Cairns used autoradiography and very careful

John Cairns in an undated photo.

isolation of phage T2 DNA to determine that the length of the DNA molecule was 52 μm. From this, and knowing that there is one nucleotide per 3.4 Å and that the average molecular weight of a nucleotide is 357, he calculated that the mass of the T2 DNA molecule was 110×10^6. He concluded that there was no need to postulate "… anything other than an uncomplicated double helix as the form of the T2 DNA molecule."

When Cairns returned to Cold Spring Harbor for the 1962 Symposium, he had no intention of running the place. He was being courted by Harvard Medical School's Department of Microbiology, and the Symposium would give him a chance to visit Boston as well. To his surprise, however, when he arrived at Cold Spring Harbor, Paul Margolin, by then a senior staff member of the Laboratory (indeed, one of the only staff members), asked him whether he would be interested in becoming its director. Cairns was flattered, but taken aback; administration was not his career goal. He asked his former supervisor Al Hershey what he thought, and Hershey, characteristically, said he did not care who was director, but that no one in their right mind would do it.

Some of biology's heavyweights, however, prevailed upon Cairns; Watson, Delbrück, and Matthew Meselson all said the Lab needed him (Cairns later learned that they had tried and failed to hire Sydney Brenner and Norton Zinder). Ed Tatum, chairman of the Laboratory's Scientific Advisory Committee, offered Cairns the directorship of the new institute that would formally combine the two Cold Spring Harbor laboratories, and Cairns accepted.

Cairns may have been the ideal candidate. He appreciated the great tradition of Cold Spring Harbor and its place among scientific institutions of research and training. He was also aware of the difficulty of financing an independent laboratory if it had no endowment for simple maintenance of facilities and long-term support for salaries and pensions, no matter how abundant the grants for research fellowships, symposia, and courses. To Cairns, Cold Spring Harbor was "Arcadia," a place with almost mythical connotations, a scientific idyll. Such imagination and romance on Cairns' part may have been a prerequisite to tackling the job of restoring the institution to its former status. He had, in short, the motivation and idealism to try to save Cold Spring Harbor and the brains to figure out how to do it.

On July 1, 1963, Cairns assumed his new office as Director of the CSHL. There was now one institute at Cold Spring Harbor, overseen by trustees from the eight universities, LIBA, and the Wawapex Society. Ironically, the realization of Demerec's dream to unite the two laboratories came as a last-ditch effort to avert collapse.

CAIRNS SHOWS DNA REPLICATION IN ACTION AND HINTS AT MORE DNA POLYMERASES

Summary

John Cairns provided autoradiographic evidence that a phage chromosome is a single-stranded DNA molecule and that the Kornberg DNA polymerase repaired, rather than replicated, DNA.

The Discovery/The Research

In 1961, Cairns spent a year in Al Hershey's laboratory where he intended to measure the length of the bacteriophage T2 DNA molecule and, using the average mass of a nucleotide, calculate the mass of the T2 DNA molecule and compare it with the mass determined by other methods. Electron microscopy was not suitable because the DNA molecule would not fit on the specimen grid, so Cairns turned to autoradiography using tritiated thymidine to label the DNA. Using techniques developed by Joseph Mandell and Hershey, Cairns isolated long T2 DNA molecules but found it difficult to spread the labeled molecules on a slide so that they remained fully elongated. In the end he was able to find a sufficient number of molecules to measure their length—52 μm. Using 357 as the average mass of a nucleotide, Cairns calculated the mass to be 110×10^6, similar to accepted values. He concluded that the T2 DNA molecule was "... an uncomplicated double helix."

Cairns second contribution to research at CSHL concerned DNA replication (see p. 248). By 1958, Arthur Kornberg had isolated and purified a polymerase that he believed was responsible for DNA replication. However, there were hints that the Kornberg enzyme was not responsible for DNA replication. Most notably, the enzyme had 5′-exonucleolytic activity; why would an enzyme that synthesized DNA also be able to destroy it? Cairns wondered if the Kornberg enzyme was responsible for excision and repair activities and not replication. If this was the case, it should be possible to find an *E. coli* mutant that could still grow even without the Kornberg DNA polymerase.

Cairns first developed an assay for the enzyme that could be used to test hundreds, if not thousands, of potential mutant bacteria. He and his research assistant Paula De Lucia treated cells with a powerful mutagen, plated out the individual cells, picked colonies, and grew them overnight. They made extracts of the cells and tested them for polymerase activity. After testing 3477 colonies they found one, p3478, with only 0.5%–1.0% of normal enzyme activity, but the same growth properties as normal cells. However, p3478 was defective in DNA repair—it was sensitive to UV light and could not grow in low concentrations

John Cairns at the 1978 Symposium.

Continued

of methylmethanesulfonate (MMS), a carcinogen that did not affect normal cells. When MMS-resistant cells arose, it was found that they had regained their polymerase activity.

Julian and Marilyn Gross, also at Cold Spring Harbor, worked out the genetic analysis of p3478 and mapped the mutation to a gene that was named *polA* (pronounced "Paula"). In the next few years, two more DNA polymerases were found by Kornberg's son, Thomas: Pol II (responsible for DNA repair) and Pol III (the main polymerase for DNA replication in *E. coli*).

Significance

Cairns' autoradiographic measurements of the T2 chromosome confirmed what everyone expected—that it was a double helix. His and De Lucia's demonstration that the Kornberg enzyme was not responsible for replication led to the discovery of two additional DNA polymerases and to research on the complex machinery of DNA replication in both prokaryotes and eukaryotes.

The Scientist

John Cairns was born on November 21, 1922, in Oxford, England. He earned his B.A. degree in 1943 and his medical degree in 1946 at the University of Oxford. He spent the years 1950–1963 in Australia at the Walter and Eliza Hall Institute in Melbourne. Cairns spent 1957 as a research fellow in Renato Dulbecco's lab at Caltech and in Al Hershey's laboratory from 1960 to 1961. Between 1963 and 1971, he was at CSHL, as Director from 1963 to 1968. In 1971 he went to the Imperial Cancer Research Fund, and from 1980 to 1991 was at the Harvard School of Public Health. Cairns was made a Fellow of the Royal Society in 1974 and in 1981 a MacArthur Foundation Fellow.

Paula De Lucia joined Cairn's laboratory in 1964 and stayed at CSHL until 1971 when she moved to Washington State.

—J.C.

13

Troubles Continue

Although there was a new institute and a new director, change was slow and the 1960s, like the 1950s, were dominated by concerns over the survival of the Cold Spring Harbor Laboratory for Quantitative Biology (CSHL) rather than the research going on in the laboratories. It was fortunate that during this period the Symposia were both intellectually and financially successful. Indeed, the educational program at the Laboratory continued to be an island of stability and excitement in a period of continuing uncertainty.

John Cairns Faces a Financial Crisis

Cairns was rapidly disabused of his view of Cold Spring Harbor as "Arcadia." As he wrote some years later:

> It was a strange sensation to find myself, after a 12,000 mile journey to the richest country in the world, director of an institute that was at once world-famous and yet decrepit beyond belief, busily engaged in subsidizing its multifarious summer program and yet virtually bankrupt, possessed of devoted alumni scattered about the globe and yet newly under the guidance of a predominantly scientific board whose members were mostly not drawn from these alumni.

Although the budget for the year ending April 30, 1963 (i.e., that preceding Cairns's directorship) showed total receipts of somewhat more than $400,000 and expenditures of nearly $384,000, this had been achieved by neglecting maintenance. In a memorandum dated

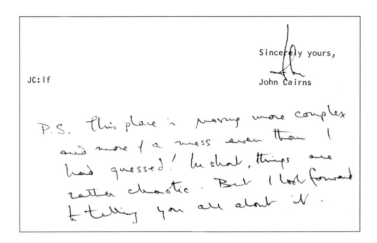

John Cairns's comments on the state of CSHL in a letter he wrote to Jim Watson, dated June 26, 1963. "This place is proving more complex and more of a mess even than I had guessed!"

July 29, 1963, Cairns described the dire straits of the CSHL finances (his emphasis):

If the Laboratory disregards the accumulated problems of maintenance it can survive about one year. If it attempts to maintain its buildings in the manner to which they should have been accustomed, it will go out of business almost at once.

There were some encouraging signs. For one, a site survey, funded by a grant from the National Science Foundation, revealed that to return all the buildings to a sound condition would cost about $750,000. Although a large sum, this was only one-fourth the assessed value of the buildings. The Board of Trustees were apparently determined to develop a long-term strategic plan for the Laboratory that would consolidate the facilities of the now-united laboratories and to launch a fund-raising campaign. Several companies signed on as corporate sponsors of the Laboratory with pledges of support and 88 "Friends of the Laboratory," mostly scientists who had been closely connected with the Laboratory, pledged to give at least $50 each. More than $50,000 came in gifts and donations from LIBA.

CSHL did well in recruiting research funds. From about $280,000 per annum in research grants in 1963–1964 and 1964–1965, this figure jumped to nearly $375,000 in 1965–1966, with funds coming from federal sources and foundations. By keeping operating expenses flat, in 1964 CSHL's unobligated cash reserve surpassed expenditures, and by the next year Cairns had managed to build a small reserve fund. The unobligated cash reserves topped $100,000 by mid-1965—about 2 months of average disbursements.

Cairns squeezed out about $25,000 per annum from the Laboratory's budget to increase maintenance and repair. He gave first priority to the restoration of the "most valuable and least dilapidated" structures, and over 4 years at a total cost of about $100,000, he put approximately half of all the Laboratory's buildings into good condition. The trouble was, it would take about five times that sum to repair the other half of the structures, which were in worse shape. In another 4 years, Cairns thought, the Laboratory might reach this goal—if there was no further

```
                    2(a)  TOTAL CAPITAL REQUIREMENTS


Required with the greatest urgency
         1.  Repair or replacement of sewage leaching field    $20,000
         2.  Capital for the accounts                          100,000
         3.  Residual debt to builder and architect             12,331
         4.  Old Grants - Dr. Abramson (Macy)                    4,005
                                        Total                 $136,336    $136,336

Required with some urgency

         5.  Purchase of land from Mrs. DeTomasi                60,000
         6.  General repairs (from 1a)                          60,000
         7.  Clearing and tidying grounds                       20,000
                                                              140,000      140,000

Required in the next two years

         8.  Renovate Animal House                             100,000
         9.  Replace or renovate Osterhout Cottage             20,000
        10.  Construction of six married quarters              90,000
        11.  Renovate Jones Lab for Nature Courses             20,000
        12.  Replace Dormitory with DeForest Apt.              50,000
        13.  Renovate and re-equip workshop                    30,000
                                                              310,000      310,000
                                                                         $586,336
```

Capital expenditures ranked by their urgency, from a report prepared by John Cairns, dated July 29, 1963.

deterioration in the meantime. Cairns himself and his family carried out minor maintenance. If Demerec's habitual wandering into buildings and turning off lights was a trademark of the previous administration, one for this period is the Cairns family cutting the grass of the Laboratory lawns.

These economies helped Cairns maintain the science program and chip away at the mountain of needed maintenance. But they were mere fingers in the dike; the only way to lower the floodwaters was to get an endowment. Demerec had bemoaned the lack of an endowment, but Cairns now needed one more than ever. Unfortunately, recognition of the Laboratory's contributions to the world of biology did not help much. With characteristic vividness, Cairns posed the problem:

> After all, it is only by being able to draw upon some independent funds for some part of each staff member's salary that the Laboratory can consider itself, in any proper sense, the employer of scientists. Without endowment, the staff will inevitably dwindle. And without staff, the Laboratory will cease to exist. It would be fitting, therefore, to be able to say that some start had been made in this quest for endowment. There is, however, no such news to report nor, apparently, even the prospect of it.

Research

The decidedly sad state of the research program Cairns inherited was evident from the first report bearing his name as director. Covering the period 1962–1963, he could list only two research groups at the CSHL, those of Margolin and Umbarger. Umbarger was active with six researchers, but Margolin listed only one. In 1964, there was one additional research group, Cairns's own, with one assistant, Cedric Davern.

With scant budgets, Cairns did his best to maintain the scientific staff, but it was difficult. Umbarger left for Purdue and Cairns replaced him with bacterial geneticist Joseph Speyer in the fall of 1964. In 1965,

Peter Lengyel came as a visiting investigator, bringing the total number of research groups to four (Margolin, Cairns, Speyer, and Lengyel). Hershey and McClintock were not strictly part of CSHL, although Hershey contributed to the CSHL annual report. McClintock did not; all her reports appeared in the Carnegie's Yearbook.

It must be said that, with some exceptions, research in CSHL was not exciting through the 1960s. Paul Margolin's group focused on mutations, gene transcription, and gene expression in *Salmonella*. In 1965, Ronald Bauerle in Margolin's lab reported that the gene cluster for the tryptophan pathway consisted of five genes, apparently grouped into two independent functional units, and by 1966, it was clear that the tryptophan operon differed from the better-known operons such as the lactose operon. Speyer, who had worked with Severo Ochoa using synthetic polynucleotides to decipher the genetic code, used mutagenesis to examine the copying of DNA and the reading of genetic information into RNA and protein.

In spite of all his administrative nightmares, Cairns completed an interesting experiment in this period. On his return to Australia following his year in Hershey's laboratory, Cairns had used autoradiography

John Cairns (right) *and his son William with Matt Meselson.*

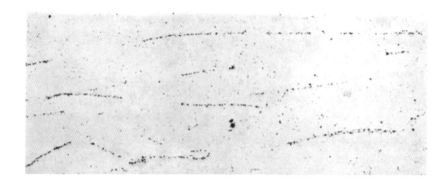

HeLa cell DNA after labeling with tritiated thymidine for 3 hours. Stretches of labeled DNA are 50–100 μm long.

Elizabeth Burgi at work in Al Hershey's laboratory.

to show that the *Escherichia coli* chromosome is circular and that during replication both new strands are labeled with ³H-thymidine. Cairns also set up a similar experiment using HeLa cells but did not examine the autoradiographs before returning to Cold Spring Harbor. Some 3 years later, Matt Meselson phoned to propose a collaboration to examine DNA replication in *Drosophila* chromosomes. Cairns remembered that the HeLa cell autoradiographs were somewhere in a drawer. He found and developed them and saw the short replicons typical of eukaryotic DNA replication.

With the continuing backing of the Carnegie, Hershey's group in the Genetics Unit flourished. Perhaps the most interesting results were from studies of phage λ. With Elizabeth Burgi and Laura Ingraham, Hershey found that DNA molecules isolated from phage λ aggregated at high concentration and that these aggregates could be broken by spinning a flat blade through the solution. By studying how these aggregates broke down, Hershey proposed that each phage λ chromosome had two complementary cohesive sites far apart on the chromosome. Aggregates were formed by end-to-end joining of molecules through these sites. In 1971, Ray Wu and Ellen Taylor sequenced the cohesive ends and showed that they were only 12 nucleotides long. cos sites (as the cohesive ends came to be called in the 1970s) became important with the advent of recombinant DNA techniques when phage λ was adopted for DNA cloning.

Anna Marie Skalka joined Hershey's group in 1964 and showed that transcription of the phage chromosome started in the GC-poor right half of the chromosome and moved leftward, accelerating as it went. With Harrison Echols, visiting from the University of Wisconsin, Skalka identified two λ genes involved in controlling transcription, the first initiated transcription of genes involved in the lytic cycle, and the second gene appeared to control transcription of the late genes.

Barbara McClintock continued to work on transposable elements in maize, uncovering seemingly ever more baroque phenomena. She received little encouragement but was immediately drawn to Jacob and Monod's operon model that she heard in June at the 1961 Symposium. Just 3 months later, McClintock published an article making explicit parallels as she saw them between the regulator, operator, and structural genes system proposed by Jacob and Monod for bacteria and the *Spm* system that she had studied. There was also evidence that the phenomena that she was studying in maize had wider significance. For example, the bacteriophage Mu was found to insert its chromosome into the *E. coli* chromosome and the behavior of unstable loci in *Drosophila* could be accounted for by transposable elements, as could strange mutations in the bacterial *gal* operon. All this was a prelude to an increasing realization that genomes were dynamic.

Anna Marie Skalka at CSHL, 1967.

The Symposia Prove Essential for the Continuing Existence of CSHL

Fortunately, despite the difficulties Cairns was experiencing, the Symposia during the 1960s were at the cutting edge of the emerging field of molecular biology and genetics, and they track the emergence of that field in the period before recombinant DNA changed everything. The Symposia volumes were required reading in a period when there were few journals publishing papers in molecular biology, and through these volumes the Laboratory became associated with advances in molecular biology even if CSHL contributions to research in this field were rather small. In addition to their financial importance, these Symposia demonstrated to the world that CSHL was a going concern.

The 1963 Symposium, the first of Cairns's directorship although organized by Umbarger, was on the *Synthesis and Structure of Macromolecules* and reflected the contemporary interest in the flow of information from genes to RNA to proteins. The science was still at a stage where the entire sequence of events could be discussed in a single meeting: synthesis and structure of DNA, transcription, translation, protein synthesis, and the genetic code. Indeed, there was a remarkable mixture of molecular and genetics approaches to the topic. Out of these studies was emerging an increasingly coherent picture of how genetic information is stored in the DNA, read off by polymerases to form messenger RNA, and conveyed by that mRNA to the ribosome, where the nucleic acid code was translated into a protein molecule.

The most significant session was the last, which included updates from the laboratories of Nirenberg and Ochoa on the genetic code. The first triplet had been reported in 1961 by Matthaei and Nirenberg using a synthetic messenger RNA containing nothing but uracil. They showed

Joseph Speyer and Marshall Nirenberg converse at the 1963 Symposium on Synthesis and Structure of Macromolecules.

that it produced a string of the amino acid phenylalanine. Then, it was shown that a string of cytosines produced a chain of proline. After that, it was a race principally between Nirenberg's and Gobind Khorana's laboratories to produce new triplet combinations, named codons by Sydney Brenner. By 1963, some 50 of the possible 64 codons had been assigned to a total of 20 amino acids. Ochoa ended his presentation with an interesting speculation on the universality of the code. To complete the code took another 3 years and produced the 1966 Symposium, arguably the highlight of the Cairns years.

The 1966 Symposium, The Genetic Code, began with Francis Crick declaring "This is an historic occasion." In a wide-ranging overview, Crick described the various clever but wrong hypotheses about the code that had been proposed (some by him) over the years. But now the deciphering of the genetic code was essentially complete, and Crick presented a table showing all the codon assignments except UGA. (In 1967, Brenner, Crick, and colleagues found using genetic analysis that UGA is a stop codon.) Cairns noted that the Symposium had some 350 participants, the largest number to that date and almost 90 presentations. The Symposium volume was 762 pages long. It was, however, not the scientifically most stimulating of Cold Spring Harbor Symposia. Rather it summarized and brought to a close a grand enterprise, one that Cairns called "without parallel in the history of biology."

Cairns's Foreword to the 1969 Symposium rings true for the last three Symposia he organized and for many of those to come:

> A science comes of age when the principles on which it was founded have been vindicated and are replaced as an occupation by the accumulation of detail…. Molecular biology is now entering the stage of detail. However, the detail is so formidable and the techniques for disclosing it are so powerful that successive symposia have had to be restricted to smaller and smaller sectors of the field if they are to remain even partly digestible by the participants.

François Jacob and Herman Lewis talk at the back of Blackford. The Blackford balcony and (at the **right***) the stairs leading up to it were lost in the extension of 1990 (page 306).*

Thus, the 1968 Symposium covered only *Replication of DNA in Microorganisms*, in 1969 the topic was *The Mechanism of Protein Synthesis*, and in 1970, *Transcription of Genetic Material*. A striking feature of these meetings is the almost total absence of any genetic analysis. This was biochemistry applied to working out the mechanics of the cell, and its tools were radioisotopes, column chromatography, and the ultracentrifuge.

SUNDAY, JUNE 7th, Evening

POPULATION STUDIES IV.

E. GOLDSCHMIDT, Hebrew University, Israel: *Estimating the Rate and Outcome of Inter-ethnic Mixture among the Communities of Israel.*

C. GAJDUSEK, National Institutes of Health, Maryland: *Factors Governing the Genetics of Primitive Human Populations.*

A. C. ALLISON, National Institute for Medical Research, London: *Natural Selection in Man.*

E. GIBLETT, King County Blood Bank, Seattle, Washington: *Haptoglobin Phenotypes: Geographic Distribution and Variant Patterns.*

MONDAY, JUNE 8th, Morning

GENETICS OF SOMATIC CELLS AND CELLS IN CULTURE I.

W. SZYBALSKI, University of Wisconsin, Madison: *Chemical Reactivity of Chromosomal DNA as Related to Mutagenicity Studies with Human Cell Lines.*

B. EPHRUSSI, Western Reserve University, Ohio: *Some Properties of Hybrids between Somatic Cells.*

T. T. PUCK, University of Colorado, Denver: Title to be announced.

P. ABEL, Universität zu Köln, Germany: *The Universality of the Genetic Code.*

MONDAY, JUNE 8th, Afternoon
PICNIC

The first 2 days of the 1964 Symposium on Human Genetics. *The pace was leisurely—only four talks on Monday morning and the whole of Monday afternoon devoted to the picnic.*

Three Symposia in this period are notable as summaries of research in fields that were to be revolutionized in the 1970s by the advent of recombinant DNA techniques and DNA sequencing. The topic for 1964 Symposium was *Human Genetics*. This seems rather out of character—human genetics had not been a focus of research at Cold Spring Harbor since Davenport's eugenics a quarter of a century earlier. The program was an uncomfortable mix of population genetics, studies using cell culture to explore the genetics of somatic cells, and data from physical studies of human proteins. The symposium was not a success—attendance was low—but even so it was significant as an acknowledgment of the legitimacy of human genetics. It would be 22 years before human genetics was again the topic of a symposium, remarkably at a time when it was possible to discuss sequencing the entire human genome.

The second of these Symposia was *Antibodies* in 1967. Immunology was on a much firmer molecular footing. Nils Jerne in his summary referred to the "… tortuous road the immunology followed until its molecular elements came in to better focus." There was one topic that would have to await recombinant DNA techniques for its resolution.

The scene in Bush Auditorium during the 1964 Human Genetics *Symposium.*

Antibody variability was the subject of a remarkable discussion involving Crick, Oliver Smithies, Lee Hood, Gerald Edelman, and Jerne, but facts, not words, were needed. Facts were appearing by the time of the 1976 Symposium on *Origins of Lymphocyte Diversity,* as Susumu Tonegawa began to unravel the mysteries of antibody diversity.

Herman Slatis, Luigi Cavalli-Sforza, Walter Bodmer, and Arno Motulsky eat in Blackford Hall during the 1964 Symposium on Human Genetics.

Macfarlane Burnet and Baruj Benacerraf sit in Bush Auditorium during a break in the 1967 symposium on **Antibodies.**

The 1965 Symposium on *Sensory Perception* covered topics that had not been considered at Cold Spring Harbor since the ending of the biophysics program some 30 years earlier. The moving force behind it was Max Delbrück, who had abandoned phage genetics for a new challenge: How do external stimuli spark an organism's cells so that a chain of events is initiated that leads to a perception? Delbrück settled on the phototropic behavior of the mold *Phycomyces* hoping that it would prove fundamental to all vision. (It did not.) The meeting covered mechanoreceptors, hearing, olfactory receptors, electrical and chemical receptors, and photoreceptors, but like *Human Genetics,* this was not an intellectually successful meeting.

Nevertheless, annual sales of the 1960s Symposia volumes averaged $100,000, reaching the astonishing figure of $204,366 in 1968. As Cairns wrote many years later, the continuing existence of the Laboratory was made possible by the sales of the 1963 volume.

Cairns Resigns

Despite the worldwide high profile of CSHL through its meetings and courses programs, it was not appreciated closer to home. At this time, the costs of CSHL research, including the salaries and operating and

A coffee break at the 1964 Symposium.

Ed Tatum at his microscope, 1958.

maintenance costs, were supported by grants from the NIH and NSF. The continuing existence of the Laboratory depended on these grants brought in by the researchers, but without job security the scientists could not be expected to stay when better offers came. The loss in 1967 of Margolin, Speyer, and Davern meant the loss of research grants totaling some $280,000. Davern was replaced by Ray Gesteland, but new appointees, junior in experience, could not be expected to bring sizable grants with them. The momentum Cairns had built drained away. By the end of 1967, the only year-round research groups were those of Hershey, McClintock, Cairns, and Gesteland, although strictly speaking the Genetics Unit housing Hershey and McClintock belonged to Carnegie.

Cairns grew increasingly despondent. Not only were the scientific staff leaving, Cairns felt that the trustees were not committed to CSHL and had never provided the support he needed. This was exemplified by Ed Tatum, President of Rockefeller University and chairman of the trustees, who insisted that the board meetings be held in New York City and not at Cold Spring Harbor; thus, the trustees did not see how bad things were at the Laboratory. Cairns also felt that Tatum had misled him at

> I would also like to make it known to the trustees that, even given a marked improvement in the long term financial position, the job of director is likely to require more time than is compatible with my desire to remain a scientist. It is my hope, therefore, that within the next several years I can move into a staff position where my primary responsibility is to do science. At the same time, I would like to emphasize my willingness to help the new director, should he so desire it, in dealing with the multitude of tasks which I know all too well he must face.

John Cairns writes to Ed Tatum saying that he intends to step down as director and return to research.

the time he was being recruited by concealing the miserable financial condition of the Laboratory.

On September 9, 1966, Cairns wrote to Tatum telling him that "within the next several years," Cairns intended to step down as director and return to full-time research. He did not stipulate a date because "… this might lead the trustees to feel that their most urgent task was to find a director, instead of first focusing on the financial problems which must face the new director."

The minutes of the Board of Trustees in this period show them grappling with three challenges. The first was the perennial issue of raising money. Arthur Pardee suggested approaching the Associated Universities, but this suggestion, as before, went nowhere. Jim Watson, who had joined the board in 1965, was particularly active, approaching foundations such as Sloan, Ford, Hartford, and Markle seeking support. Bentley Glass brought in SUNY Stony Brook as a participating institute, and its contribution of $25,000 encouraged the Trustees to require the other participating institutes to provide their promised monies.

The second challenge was to provide a means for Cairns to remain at Cold Spring Harbor as an investigator. The best solution seemed to be an American Cancer Society Lifetime Research Professorship, but this required Cairns to be associated with an academic institution. Tatum was asked to explore the possibility of an adjunct professorship at Rockefeller University, but the best he could do was an "affiliate" appointment. Cairns's bitterness can be read in the letter

A pensive John Cairns at the 1965 Symposium.

he wrote to Tatum on March 1, 1967, copied to the executive committee of the trustees:

> On the one hand, the offer from Rockefeller University is so small a gesture that, from my point of view (and the Laboratory's as well), it might as well have not been made, for it carries with it no sense of responsible authority nor concept of duration. On the other hand, a five year grant ... would merely extend my tenureless status to the unprecedented age of 50—at which point I would presumably have to enter the market once again. I cannot believe that any Trustee will consider me unreasonable in rejecting this offer.

The third challenge was to find a new director and to chart a future for the Laboratory. At the October 7, 1966, executive session meeting of the Board of Trustees, a temporary long-term planning committee was formed, chaired by Watson, with Tatum, Harry Eagle, and Irwin Clyde Gunsalus as members. Watson undertook to prepare a document describing the history of the Laboratory, its influence, and future plans that could be used to seek endowment.

Watson submitted a 12-page development plan in January 1967. It was a remarkable document in which Watson set out to answer, among others, a key question: "Isn't the Cold Spring Harbor Lab a very

```
                          New and substantial finan-
cial aid is necessary.  So imperative in fact are additional
sources of support that at first we should ask: "Is it really
worthwhile to keep the Cold Spring Harbor Lab going?  Might
not its functions be taken over by the universities that em-
phasize molecular biology?  Couldn't they teach the summer
courses, or, if not, couldn't a larger, more stable summer lab
like the Marine Biological Laboratory (MBL) at Woods Hole do
the job just as well?  Isn't the Cold Spring Harbor Lab a very
expensive luxury, not a necessity meriting the continuing
attention and support of governmental agencies and private
foundations?"
```

Watson asks hard questions in his 1967 A Development Plan to Support the Cold Spring Harbor Laboratory of Quantitative Biology.

expensive luxury, not a necessity meriting the continuing attention and support of governmental agencies and private foundations?" Not surprisingly Watson's answer was "no":

> We conclude not only that a lab like CSH must continue to exist but that the CSH site and facilities now offer potentialities unlikely to be equaled at a reasonable cost at any other location in the United States.

Watson was unflinching in setting out the severe obstacles the Laboratory faced: no endowment, a too small research staff, lack of housing for summer visitors and junior staff, and only one building fit for year-round research. He followed this with some general principles governing his development plan and then set out in detail with budgets what he felt was needed. The bill came to $3,915,000, at a time when the total assets of the Laboratory were $750,332.

The Board of Trustees met in June 28, 1967. In a closed session not recorded in the official minutes, the committee debated long and hotly over the Laboratory's course and who should be Scientific Director. Watson defended the recommendations in the development plan he had presented in January. With characteristic emphasis and certainty, Watson outlined a plan of action. When he had finished, there was a moment of tense silence. Then someone, possibly Norton Zinder, said bluntly, "Jim, if you know just exactly what ought to be done, why don't you take it on?" Watson gave a characteristic pause, and then, in a musing voice, responded, "Maybe I will."

On October 17, Watson wrote to Glass with a formal proposal remarkable in its boldness, sweep, and diplomacy. He would accept the directorship of the Laboratory at the beginning of 1968, while maintaining his post on the Harvard Faculty and his joint laboratory there with Walter Gilbert. He would spend about two-thirds of each year at Harvard and one-third at Cold Spring Harbor. So far as possible, he would spend part of every month at Cold Spring Harbor. No later

Dr. Bentley Glass
State University of New York
Stony Brook, New York

October 17, 1967

Dear Bentley:

I wonder whether the following proposal might solve our current Cold Spring Harbor dilemma.

1) I be appointed the Director of the Lab beginning January 1, 1968. At the same time, I would remain a member of the Harvard Faculty, maintaining my joint lab with Wally Gilbert.

2) I would propose to spend about two-thirds of each year at Harvard, the remaining time at Cold Spring Harbor. INsofar as possible I would plan to spend part of most months at Cold Spring Harbor, the one-third time certainly not to be thought of as largely for the summer.

3) Beginning no later than September 1969, I would maintain a lab at Cold Spring Harbor, the exact space depending upon my success in finding funds to renovate quickly both James and the Animal House.

7) I see my primary responsibility as Director as the building up of the scientific staff, the finding of short-term funds to give them modern first-rate laboratories, and the acquisition of sufficient resources, so that within five years or so, the Lab sould have the resources to pay the salary of the Director.

8) In taking over these duties, I would give a guarantee of remaining for five years. Taking over for a shorter period would leave too much uncertainty. A commitment for a longer period would probably not make sense --on either side.

Watson writes to Bentley Glass proposing that Watson takes over as director of CSHL and describing what he sees as his responsibilities.

than September 1969, he would establish a laboratory at Cold Spring Harbor, the space to depend on finding the funds to renovate both James Laboratory and the Animal House. Cairns would be asked to remain as Associate Director until Watson could find an Administrative Director and to remain in charge of the annual Symposium and its editing. Ray Gesteland would be asked to serve as Assistant Director in charge of the summer programs. Watson's primary responsibility would be to build up the scientific staff, to find short-term funds to improve the laboratory facilities, and to acquire sufficient resources to permit payment of the Director's salary. Finally, he would promise to remain for 5 years, because a shorter period would leave too much uncertainty and a longer period would not be sensible for either him or CSHL. The Laboratory would also supply a rent-free residence and would cover his travel expenses between Cambridge and Cold Spring Harbor.

Watson proposed that Harvard assign one-third of his salary to the Cold Spring Harbor Laboratory in compensation for his duties there as Director of the Laboratory. He would not reduce his teaching or committee responsibilities at Harvard, and CSHL would assume part of his summer salary. In addition, Harvard would become a participating institution in support of the Cold Spring Harbor Laboratory and agreed to contribute $20,000 (not $25,000) over 5 years.

On December 10, the Executive Committee of the Board unanimously approved Watson's proposal for recommendation to the Board. Cairns was not altogether happy about the outcome. If after September 1, 1968, he was to get one-half to one-third of his salary from his administrative supervision of the symposia, editing, and publication, the remainder was left too nebulous to count on. His research grant from NSF remained to be renewed and even so would pay only one-fourth of his salary.

At the full meeting of the Board of Trustees on December 16, Watson's plan was presented with the endorsement of the Executive Committee and was fully discussed. Watson estimated that the full cost of his services to the Laboratory would not exceed $6000 annually, including travel and housing. After he had retired from the room, and after some further discussion, the Board unanimously approved the 5-year appointment, beginning February 1, 1968. The Board also passed a motion authorizing Watson to take immediate steps to recruit an administrative director. Cairns was asked to become Associate Director from February 1 until September 1, 1968, at an annual salary approximately equivalent to what he had been receiving.

In his final annual report, in 1967, Cairns eloquently and poignantly analyzed the situation facing the Laboratory, the events that led to it, what he had done to alleviate it, and the Laboratory's future prospects. His reforms and economies had brought the Laboratory into stable condition: a balanced operating budget, a small cash reserve, and buildings that, although in need of repair, were clearly worth much

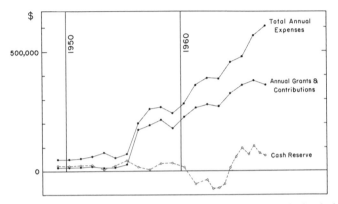

The graph that Cairns included in his last director's report in 1967. "The level of the Total Annual Expenses is a rough index of the magnitude of the operation. In the years when the cash reserve does not change, the difference between Grants & Contributions and Total Expenses is made up by income from what might be called the Laboratory's hotel and publishing business."

more than the cost of renovations. The Symposium was stronger and an extremely valuable asset, bringing scientists to Cold Spring Harbor and through the published volumes adding substantially to the Laboratory's income.

In 1968, Watson in his first annual report as director praised Cairns for his work:

> To start with, let me express the strong debt which everyone connected with Cold Spring Harbor owes to John Cairns. When in June, 1963 he came here as director, he inherited a terrible mess.... Many outsiders thus thought his task insuperable, and that Cold Spring Harbor would soon cease to exist. But with much devotion, intelligence, and great indulgence of his free time, he again made the Laboratory a going concern. Without his achievement, I would have never taken over as director.

14

1970s Research

Simply appointing a new director was not going to change things quickly, and although Cairns had, as he put it, "staunched the flow," Watson still faced the same problems: The scientific staff was depleted, the finances shaky, and the infrastructure in urgent need of care. Watson, however, took command of the Laboratory at what was, with hindsight, an opportune moment. He was able to take advantage of a significant increase in federal support of biomedical research, particularly for cancer, that occurred in his first years as director. This increased funding, together with support from foundations, not only supported research but provided moneys for renovation of buildings and laboratories. And most important of all, he was able to recruit a coterie of exceptional young scientists who rapidly made the Laboratory a center for research on molecular and cellular biology. These factors acted synergistically—the excellence of the research attracted more funding, which supported more research, which in turn attracted bright young scientists. Watson's success resulted from a combination of the power and flexibility of the resources Cold Spring Harbor offered and his skill and judgment in using them and in recognizing talented scientists and productive new research areas. He was aided by a measure of unanticipated good fortune so that by the mid-1970s Cold Spring Harbor was clearly in the ascendant.

Jim Watson in the summer of 1968 just after becoming Director of CSHL.

Al Hershey Wins a Nobel Prize

It must have been encouraging for all at Cold Spring Harbor when in 1969, at a time when the future of the Laboratory was still uncertain,

Jill and Al Hershey dancing at the 1969 Nobel Prize celebrations.

Al Hershey was awarded the Nobel Prize in Physiology or Medicine. Appropriately, he shared the prize with his two longtime colleagues in phage biology—Max Delbrück and Salvador Luria. The three had established the field of phage molecular biology, which had led to studies of other viruses and to the first steps in understanding the regulation of gene expression. By 1969, all three had moved on to other fields. Delbrück had been working on *Phycomyces* for many years, while Luria's interests had turned to a study of bacterial colicins. Hershey had published his last experimental paper in 1968 and was to retire in 1971 at age 63.

Tumor Viruses Were to Be the Key to Unlock the Mysteries of the Cancer Cell

It is not surprising that the first research program initiated by Watson at CSHL was on cancer. The final chapter of his classic textbook *Molecular Biology of the Gene* (1965), the first textbook to present the new world of molecular biology to undergraduates, was "A Geneticist's View of Cancer." The chapter's inclusion was justified, Watson wrote, in part because of "… some recent spectacular results on the induction of tumors by viruses." Tumor viruses contained few genes, and understanding how these and their protein products changed a normal cell into a cancer cell would, he was sure, provide crucial insights into the general characteristics of cancer cells.

Tumor viruses had not been a popular line of cancer research. Peyton Rous's eponymous Rous sarcoma virus discovered in 1911 and Richard Shope's rabbit papilloma virus (1933) were regarded as outliers, and the National Cancer Institute (NCI), founded in 1937, was uninterested in viruses as a cause of cancer. However, in 1953, Ludwik Gross demonstrated that leukemia could be transmitted between mice by a filterable agent—that is, a virus. Five years later, polyoma virus, causing solid

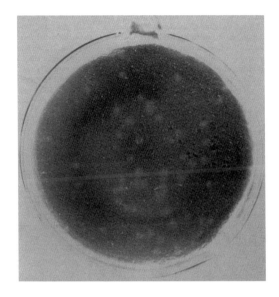

Dulbecco's plaque assay. A suspension of Western equine encephalomyelitis virus was added to a culture of chicken fibroblasts. Each of the lighter colored circles is an area of cell death resulting from the infection of a single cell by a single virus particle. The cell dies releasing virus particles, which go on to infect the neighboring cells.

tumors in rodents, was isolated, and there was an increasing enthusiasm for cancer virology. In 1965, NCI reversed its attitude and established the Special Virus Leukemia Program. In 1969, the program was renamed the Special Virus Cancer Program and finally just the Virus Cancer Program.

Renato Dulbecco was the leader in studying animal viruses. In 1952, he had developed a plaque assay using animal cells in tissue culture that made animal virus studies as convenient as research on phage. Now at the Salk Institute in La Jolla, Dulbecco was exploring the activities of cancer-causing viruses such as polyoma virus and SV40. It was a sign of the changing times that Peyton Rous, then 87 years old, shared the 1966 Nobel Prize for Physiology or Medicine for the discovery of the Rous sarcoma virus 55 years earlier.

Tumor Viruses Come to Cold Spring Harbor Laboratory

Watson's first step was to visit Dulbecco's laboratory in search of someone to lead the CSHL tumor virus program, and he found him in the

Joe Sambrook (right) *with Hatch Echols at the 1970 Symposium on* Transcription of Genetic Material. *Harrison "Hatch" Echols was an eminent λ phage researcher at the University of California, Berkeley who died at the young age of 59.*

person of Joe Sambrook. Sambrook had been working for about a year at Salk when Watson appeared, sounding him out about a move to Cold Spring Harbor. Watson suggested that Sambrook spend the summer of 1968 assisting in the Microbiology of Vertebrate Cells & Quantitative Animal Virology course taught by Gordon Sato and Philip Marcus. By the end of that summer, Sambrook was hooked, and in May 1969, he and his family moved to the East Coast.

Watson's appointment of Sambrook was a key element in the revival of the Laboratory. Sambrook, intelligent and driven, would help build and lead a remarkable group of young scientists. Interactions between Watson and Sambrook were frequently stormy but nevertheless they had, as Tom Maniatis put it, a "remarkably constructive love/hate relationship.... Most of the time their talents synergized; sometimes they mixed explosively."

Even before he had arrived Sambrook had begun work, and by the end of 1968 he had written a NIH grant application for a "Cold Spring Harbor Tumor Virus Research Center." It was an ambitious proposal requesting $1.6 million over 5 years, far larger than any previous CSHL grant.

SECTION 1

NOT FOR PUBLICATION	DEPARTMENT OF HEALTH, EDUCATION, AND WELFARE	LEAVE BLANK · [For office use only]
OR PUBLICATION	PUBLIC HEALTH SERVICE	SIE PROJECT NUMBER
REFERENCE.	RESEARCH OBJECTIVES	

ABBREVIATED TITLE OF PROJECT

Cold Spring Harbor Tumor Virus Research Center

NAME, SOCIAL SECURITY NUMBER, OFFICIAL TITLE AND DEPARTMENT OF ALL PROFESSIONAL PERSONNEL ENGAGED ON PROJECT

James D. Watson — Director of Cold Spr. Hbr. Lab.
Joseph Sambrook — Staff Investigator
Raymond F. Gesteland — Staff Investigator-Asst. Director
John Cairns — Staff Investigator
Bernhard Hirt — Staff Investigator
Lionel Crawford — Staff Investigator
Hajo Delius — Staff Investigator

NAME AND ADDRESS OF APPLICANT ORGANIZATION

Cold Spring Harbor Laboratory of Quantitative Biology
Cold Spring Harbor, L. I., N.Y. 11724

USE THIS SPACE TO MAKE A BROAD STATEMENT OF YOUR RESEARCH OBJECTIVES

We intend to study how the SV40 (polyoma) viruses replicate, hoping that our findings will be relevant to the mechanism by which these viruses induce malignant cellular transformations. Our approach will be that of the molecular biologist, visualizing SV40 (polyoma) as a very small virus whose DNA content suggests a genome of at most 8-10 genes. We hope to obtain evidence telling us what these genes do, both in the lytic cycle and when they are integrated into the host genome.

Initially much of our effort will go toward establishment of methods for large scale virus purification, using this virus both to work on the chemistry of its protein components and in the isolation of purified DNA molecules to be used in in vitro experiments in which we attempt to make biologically meaningful RNA and protein. A major effort shall also be made to isolate temperature sensitive conditional lethal mutants. We plan to use them in experiments studying how SV40 DNA, RNA, and proteins are synthesized.

Part of the 1969 NIH application for the Cold Spring Harbor Tumor Virus Research Center.

It began: "We intend to study how the SV40 viruses replicate, hoping that our findings will be relevant to the mechanism by which these viruses induce malignant cellular transformations." By March of 1969, it was approved and Watson, as he wrote to Delbrück, could breathe easier. Now he could rebuild the scientific staff and purchase new equipment while the indirect costs would help renovate the facilities.

Hiring of staff went swiftly. Watson returned to the Dulbecco laboratory and hired Carol Mulder and Heiner Westphal, brought Bernhard Hirt from the Institute for Cancer Research at Lausanne, Switzerland, and Hajo Delius from the University of Geneva. Within 2 years the number of scientists at CSHL and research groups had doubled. In addition to Gesteland's and Cairns' groups, Rudolph Werner's group was studying DNA replication in bacteriophage and bacteria, Joe Sambrook was leading the tumor virus group, and Hajo Delius was the resident electron microscopist.

The tumor virus group quickly became the Laboratory's largest research group: It had 16 members in 1970, 20 in 1971, and 27 in 1972. The group was centered in James Laboratory, the old biophysics building. In 1970, extensive renovations began on the James laboratories while a new addition to James provided offices for the tumor virus researchers, as well as for Watson. The modern accommodations suited the close-knit nature of the group. Laboratory supplies and glassware were shared between laboratories, scientists cooked meals in the kitchen at all hours, and wrestled and invented games of skill in the seminar rooms and halls.

Members of the tumor virus group—the "James Gang"—focused on the DNA viruses, polyoma and SV40, and later adenovirus. The work was fast-paced, collaborative, and exploratory. Many different projects went on simultaneously including studies of RNA transcription and translation of RNA into protein. Transcription was one of the most active areas, because so many questions seemed to open up in eukaryotic systems. Other scientists sought to compare virus-infected cells with uninfected cells. Could tumor viruses throw a cell's growth machinery into overdrive, causing it to reproduce faster or more frequently than normal? Could viral infection produce tumor-like growth in cells in Petri dishes? All research was directed at answering the questions: How did viruses cause cancer and what did that tell us about cancer cell biology?

The first significant paper from the James Laboratory was published by Sambrook, Phil Sharp, and Walter Keller in 1972. It described molecular analyses of the locations and transcription of SV40 genes and its style was typical of many papers that followed. Sambrook wrote later that the paper also typified the close collaborations in the tumor virus group, with individual members contributing their skills to a common end. In this case, Keller produced the RNA drivers needed to separate the SV40 DNA

J. Mol. Biol. (1972) **70**, 57–71

Transcription of Simian Virus 40

I. Separation of the Strands of SV40 DNA and Hybridization of the Separated Strands to RNA Extracted from Lytically Infected and Transformed Cells

JOE SAMBROOK, PHILLIP A. SHARP AND WALTER KELLER

Cold Spring Harbor Laboratory
Cold Spring Harbor, N.Y. 11724, U.S.A.

(*Received 7 April 1972*)

Asymmetric RNA was synthesized *in vitro* from SV40 component I DNA using *Escherichia coli* DNA-dependent RNA polymerase. When denatured, unit-length, single-stranded SV40 DNA was incubated in the presence of 6- to 20-fold excess of asymmetric RNA, about 50% of the DNA (E-DNA) formed DNA–RNA hybrids. The unhybridized DNA (L-DNA) was separated from the DNA–RNA hybrids by chromatography on hydroxyapatite. E-DNA and L-DNA were shown to be the complementary strands of SV40 DNA. After further purification and shearing, the separated strands were hybridized to RNA extracted at different stages of lytic infection and to RNA from transformed cells. "Early" RNA contained sequences complementary to 30% of E-strand DNA; "late" RNA bound to 30 to 35% of E-strand DNA and to 70% of L-strand DNA. RNA from SV3T3 cells hybridized with 50% of E-strand DNA and 15 to 20% of L-strand DNA.

The first of many papers to be published from the tumor virus group.

strands, Sharp prepared highly radioactive SV40 DNA, and Sambrook carried out the hybridizations and the strand separations on hydroxyapatite columns. They showed that the SV40 early expressed genes mapped to one DNA strand, whereas the late expressed genes to the other strand.

Restriction enzymes cutting DNA at specific sequences were becoming available, and Sambrook and Bill Sugden prepared two restriction enzymes from *Haemophilus parainfluenzae, HpaI and HpaII*. These enzymes cut SV40 at three sites and a single site, respectively. Now SV40 mRNAs could be hybridized to these fragments and thus mapped to the intact genome. However, purification of restriction enzymes was a slow business, as each step in the process required cumbersome assays using radioactive SV40 RNA and chromatography columns. In the course of preparing and characterizing these two enzymes, Sambrook, Sharp, and Sugden found that ethidium bromide, a mutagen because it binds to DNA,

Detection of Two Restriction Endonuclease Activities in *Haemophilus parainfluenzae* Using Analytical Agarose–Ethidium Bromide Electrophoresis†

Phillip A. Sharp,* Bill Sugden, and Joe Sambrook

ABSTRACT: A rapid assay for restriction enzymes has been developed using electrophoresis of DNA through 1.4% agarose gels in the presence of 0.5 μg/ml of ethidium bromide. The method eliminates lengthy staining and destaining procedures and resolves species of DNA which are less than 7×10^6 daltons. As little as 0.05 μg of DNA can easily be detected by direct examination of the gels in ultraviolet light. Using this technique, we have identified two different restricting activities in extracts of *Haemophilus parainfluenzae*. The two activities have different chromatographic properties on phosphocellulose and Bio-Gel A-0.5m, and they attack SV40 DNA at different sites. One activity (*Hpa* II) cleaves SV40 DNA at a single position situated 0.38 fractional genome length from the insertion point of SV40 sequences into the adenovirus SV40 hybrid Ad2++ND$_1$. The other activity (*Hpa* I) cleaves SV40 DNA at three sites which appear to coincide with 3 of the 11 cleavage points attacked by a restriction system isolated from *H. influenzae* strain *Rd*.

Title page and abstract of the paper in which Sharp, Sugden, and Sambrook describe using the combination of agarose gel electrohoresis, ethidium bromide staining, and UV illumination for examining fragments of DNA produced by restriction enzymes.

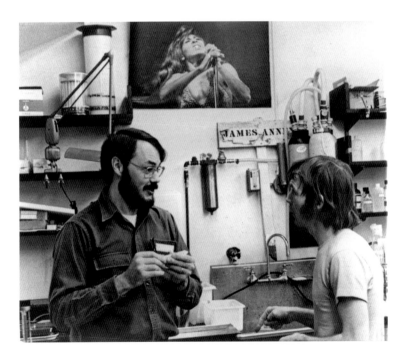

Phil Sharp, Tina Turner, and Joe Sambrook in James Laboratory.

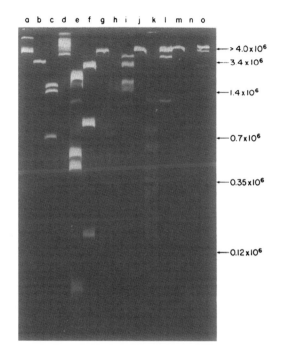

An agarose gel showing HpaI *and* HpaII *restriction enzyme digests of SV40, φR, T7, and adenovirus 2 and 3 DNAs.*

could be used to stain DNA fragments separated by electrophoresis in agarose gels. Electrophoresis in agarose gels with ethidium bromide staining had the advantages of being fast, simple, accurate, and not requiring highly radioactive SV40 DNA. One of the first to appreciate it was Herb Boyer who, with Stan Cohen, was a pioneer in developing recombinant DNA techniques. Boyer came to give a talk at CSHL and Sambrook and Sharp showed him an agarose gel with adenovirus fragments stained with ethidium bromide. Boyer later reported: "I said 'Thank you, lord!'" because it would remove the drudgery of analyzing DNA fragments. The technique rapidly became used in laboratories worldwide.

The tumor virus group published two other papers using restriction enzymes in 1972, analyzing the genomes of SV40 and adenovirus. The latter DNA virus had been introduced to Cold Spring Harbor

Laboratory by Ulf Pettersson from Sweden who arrived in Cold Spring Harbor in 1971. Adenoviruses are larger and more complex than SV40 having genomes of between 27 and 48 kb. They became the preferred subject for research in the tumor virus group, with profound consequences a few years later.

The War on Cancer and Further Expansion

Thus CSHL was well positioned to benefit from the National Cancer Act signed into law by President Nixon at the end of 1971. A substantial fraction of the vast new funds allocated to the NCI would be used to establish cancer research centers around the country. Watson applied to make Cold Spring Harbor a charter member of this elite group, an application that Sambrook recalls as "... a bit of a dog's breakfast," with rather loosely linked sections drawing on the expertise at the Laboratory. The Laboratory was awarded an NCI Program Project grant, locally named

President Nixon signs the National Cancer Act on December 23, 1971.

the DNA Tumor virus grant, which created the Cold Spring Harbor Cancer Research Center. The new grant approved the allocation to the Laboratory of ~$1 million each year for the next 5 years. This Program Project grant from the NCI became the core of the Laboratory's research support and has been renewed every 5 years since. The funds allowed the Lab to hire more scientists and to strengthen the administrative and support services. In the 5 years of the grant, Demerec Laboratory underwent a total renovation, the number of scientists almost doubled, and the total staff increased by 50%. The major expansion of cancer research enabled CSHL to apply for and receive an NCI Cancer Center Support Grant in 1987. With this recognition, CSHL was became an NCI-designated Cancer Center, and the grant supports many core research facilities. It, too, has been renewed every 5 years and is still active.

The tumor virus group continued to be the largest at the Laboratory, and although the total number remained at about 29, there was turnover in the group. Mike Botchan and Terri Grodzicker were important recruits. Botchan came to work primarily on SV40, studying the

Terri Grodzicker and Mike Botchan.

integration of SV40 DNA in the host cell genome. Grodzicker worked with Sambrook and Sharp to develop and use restriction fragment polymorphisms to map temperature-sensitive mutations in adenovirus. The research focus shifted somewhat from SV40 to adenovirus but the goals were the same—to understand, through knowledge of viral genes and their expression, the molecular changes in the transformed cell.

Another cancer-related research area used cells in tissue culture. Mammalian cell biology had been taught in the summer course beginning in 1964, but it did not become a focus of research until the arrival of Robert Pollack, Klaus and Mary Weber, and Elias Lazarides in 1972. Pollack established the Mammalian Cell Genetics group in the renovated Demerec Laboratory, focusing on what happens at the cellular level when infection by a virus "transforms" a normal cell into a cancer cell. What are the characteristic changes in the behavior of these cells? Can they revert to normal behavior? What molecular changes are involved? Pollack examined the properties of transformed cells—their growth, movement and interactions—in clones of mouse 3T3 cells transformed with SV40. Weber's laboratory used fluorescently labeled antibodies to examine the structural proteins—actin, myosin and tubulin—of normal cells and, in collaboration with Pollack, of transformed cells. Weber and Lazarides demonstrated that there was a complex arrangement of structural protein bundles in normal cells.

Klaus Weber in 1970.

Pollack also taught the Quantitative Microbiology of Animal Cells in Culture course, and Janet Mertz from Paul Berg's laboratory at Stanford was one of the students in the 1971 course. It was customary for students to give presentations of the work that they were doing in their home laboratory, and Mertz described experiments trying to use SV40 as a vector for introducing foreign genes into mammalian cells. Berg was proposing to use phage λ as the source of the foreign DNA. Pollack was dismayed at the possibility that rather than SV40 acting as the vector carrying phage λ DNA into mammalian cells, phage might act as the vector and

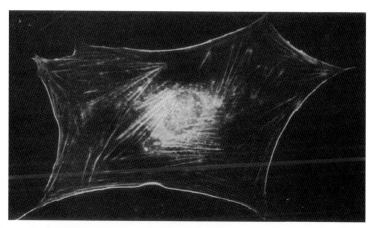

A fluorescently labeled anti-actin antibody detects fibers in a non-muscle cell (mouse 3T3 cell) from a paper by Lazarides and Weber (1974).

Bob Pollack in 1979.

carry SV40 into, for example, *Escherichia coli*. The latter is a major component of the human gut microbiota, and so SV40-infected *E. coli* could spread, carrying the tumor virus to who knew where. Pollack was so worried that he phoned Berg, who put off doing the experiments. Now alerted to the possible dangers of these experiments, Berg and others sent a letter to *Science* advocating a moratorium on gene transfer experiments. (Watson was a signatory, something he soon regretted.) From this small beginning and a subsequent conference at Asimolar, there arose a bureaucratic network of regulations administered by the Recombinant Advisory Committee (RAC). These regulations were relaxed some years later, but not before research that would have benefitted from recombinant DNA techniques had been seriously delayed.

Not all science at Cold Spring Harbor was centered on cancer, and while Hershey continued, as ever, to study phage λ in the Carnegie Genetics Unit, David Zipser began work on phage μ. He gradually rebuilt his group around this phage, hosting workshops and putting undergraduates to work on μ problems. Ahmad Bukhari, one of Zipser's postdocs, went on to establish his own laboratory at CSHL, exploring

Ahmad Bukhari in his laboratory (1973).

Rich Roberts in 1975.

phage μ biology and how it behaves like one of Barbara McClintock's transposable elements, moving about the host cell genome. Bukhari was the lead μ researcher until his untimely death of a heart attack at age 40 in 1983.

Richard Roberts arrived at about the same time, bringing his skills in nucleic acid chemistry from Harvard. His goal was to explore using restriction enzymes to sequence DNA, but soon the enzymes themselves became the focus of his research and his laboratory purified many new restriction enzymes and provided them to other laboratories inside and outside CSHL. Eventually the task became too great, and New England BioLabs, where Roberts was a member of the scientific advisory board, became the go-to source of restriction and other enzymes.

Through much work by Watson, John Cairns was awarded an American Cancer Society Professorship and returned with relief to research. There were hints that the DNA polymerase discovered by Arthur Kornberg in *E. coli* was not the DNA polymerase responsible for replication of DNA but had some other role in the cell (e.g., DNA repair).

Cairns devised and Paula De Lucia, his assistant, carried out an ambitious experiment. They would mutate *E. coli* and search for mutants that could divide but were deficient in the Kornberg enzyme. Cairns developed an in vitro assay for the enzyme suitable for screening colonies of many hundreds of mutants. It was just as well that the assay was up to the job; it was not until De Lucia tested the cells of the 3478th colony that she found a mutant with no measurable DNA polymerase activity but normal growth. The mutant was called *polA,* pronounced "Paula" by those in the know. Kornberg's son, Tom, went on to isolate DNA polymerases Pol II and Pol III; the former is involved in DNA repair, whereas the latter, in a complex with many other proteins, is the polymerase responsible for DNA replication in *E. coli.* Cairns's 1970 paper in *Nature,* however, was his last from CSHL. In 1973, he and his family moved to England, where he became Director of the Mill Hill Laboratory of the Imperial Cancer Research Fund. His interests moved to epidemiology of cancer and he wrote *Cancer: Science and Society* (1978), moving to the Harvard School of Public Health in 1980 and retiring in 1991.

Some scientists came to CSHL not as a matter of choice but of necessity. Tom Maniatis had been working on the phage λ operator

Paula De Lucia, who found the E. coli polA *mutant lacking DNA polymerase but able to replicate DNA.*

The balcony of the Cambridge city council chambers during the debate on whether there should be a moratorium on recombinant DNA experiments in the city. Tom Maniatis is in the dark shirt, left of center.

as a postdoctoral fellow in Mark Ptashne's laboratory, and he and Arg Efstratiadis had begun to develop methods to make full-length cDNAs for cloning eukaryotic genes. However, Cambridge Mayor Al Vellucci, with the backing of Science for the People, banned DNA cloning in Cambridge. (Vellucci is remembered for writing, apparently in all seriousness, to Philip Handler, President of the National Academy of Sciences, asking whether recent sightings of Bigfoot were related to recombinant DNA experiments at Harvard.) Watson offered Maniatis sanctuary, and he moved to Cold Spring Harbor in 1975. Efstratiadis prepared human globin cDNAs in Cambridge and then mailed them to Cold Spring Harbor, where Maniatis refined the steps necessary for

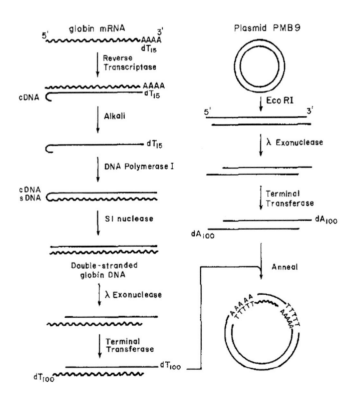

A figure from the paper by Maniatis, Kee, Efstratiadis, and Kafatos (1976) showing their procedure for cloning cDNAs.

cloning them into a plasmid. Their successful technique was the starting point for groups around the world to begin cloning genes.

A Scientific Revolution

By the early 1970s, adenovirus was the object of research for several groups at CSHL, in addition to the tumor virus group in James Laboratory. These various researchers working on adenovirus were getting results that were difficult to interpret. Ray Gesteland's group found that some viral mRNAs were longer than predicted on the basis of the proteins they coded and that mRNAs coding for the same protein were selected by DNA fragments known to be far apart in the genome. Rich Gelinas, who arrived in Roberts' laboratory in 1974, found that mRNAs known to be coded for at different parts of the genome remarkably had the same 5′ sequence. There were also the findings of Louise Chow and Tom Broker, who had arrived in 1975, taking over the Electron Microscopy group from Hajo Delius. Chow and Broker were experts in R-looping, a technique devised in Norman Davidson's laboratory, and were using this to map adenovirus late mRNAs to the viral genome. They found that many R-loops had unhybridized 5′ tails—that is, the sequences at the 5′ end of the mRNA were not matched by what should have been their DNA coding sequences.

These anomalies were resolved in 1976 by experiments performed at Cold Spring Harbor Laboratory and by Phil Sharp (by now at MIT) and Sue Berget.

At CSHL, Roberts, Gelinas, Chow, and Broker began a collaboration using single-stranded DNA restriction fragments as probes to identify the origins of the free 5′ tails that they had observed. They were able to show that sequences from three different positions in the genome were complementary to the 5′ tails of mRNA. The explanation for all the experimental results was that the coding sequence of a gene is interrupted by noncoding regions. The latter were named introns and the

Louise Chow and Tom Broker at Cold Spring Harbor Laboratory in 1974.

Cell, Vol. 12, 1–8, September 1977, Copyright © 1977 by MIT

An Amazing Sequence Arrangement at the 5′ Ends of Adenovirus 2 Messenger RNA

Louise T. Chow, Richard E. Gelinas, Thomas R. Broker and Richard J. Roberts
Cold Spring Harbor Laboratory
Cold Spring Harbor, New York 11724

The "Amazing" paper, the first of four describing RNA splicing.

coding sequences exons. The pre-mRNA transcribed from DNA is a full-length transcript of the gene sequence. This undergoes "splicing," in which the introns are cut out and the exons stitched together to form the mRNA that goes to the ribosomes for protein synthesis.

The results of this and other findings were published back to back in four CSHL papers in *Cell*. The title of the first paper was a succinct description of the results: "An Amazing Sequence at the 5′ Ends of the Adenovirus 2 Messenger RNA." Roberts recalls that Ben Lewin, editor of *Cell*, objected to the word "amazing." Roberts asked Lewin what he thought of the results, and when Lewin replied "amazing," Roberts's case was made.

It was, indeed, an amazing, unforeseen, and revolutionary finding, overthrowing the long-held assumption that the DNA sequence coding for a gene was collinear with its mRNA transcript. Roberts and Sharp shared the 1993 Nobel Prize in Physiology or Medicine.

The Robertson Gift Provides an Endowment

Although the financial position was much improved by the NIH tumor virus grants and the successes of CSHL scientists at getting grants from other sources, the Laboratory's cash reserves were low and it was living something of a hand-to-mouth existence.

What was needed was an endowment and the lack of one was a recurring lament of Cold Spring Harbor directors. Reginald Harris wrote in 1929 that, "The undertaking of permanent research calls for the establishment of an adequate endowment. We need further laboratory and living accommodations." Sixty years later Watson echoed Harris's words:

> [T]he Lab badly needs a real benefactor, but with much love it will probably survive without one. Of course, I dream an angel will appear soon and make me free of any serious worries for at least a month.

Watson was unrelenting in his search for benefactors, and in 1972 when he and Liz were in California talking to potential donors, he received a phone call from Ed Pulling, chairman of the Long Island Biological Association. Pulling told Watson that a local philanthropist was considering making a substantial gift to the Laboratory.

Charles Sammis Robertson had returned to his native Huntington after graduating from Princeton, but in the aftermath of the Depression he had had to take a job selling real estate. One day a young woman, Marie Hoffman, came to the office looking to purchase property in the area. She was a granddaughter of George Huntington Hartford, one of the founders of the Great American Tea Company, later known as A&P, which became America's largest grocery retail company. Charles and Marie married, bought parcels of land in Lloyd Harbor to create a large estate, and in 1938 built a house designed by Mott Schmidt. In 1957, the trust fund established by her grandfather was dissolved and Marie inherited 10% of the A&P stock, some $85 million.

The Robertsons made substantial gifts to Princeton but were dissatisfied (and increasingly so) with Princeton. Charles was looking to benefit a local institution, and in 1973, he gave $8.5 million to the Laboratory as an endowment for research. (Marie had died in 1972.) The annual income of some $250,000 from the endowment would for the first time provide the director with discretionary funds to be used as needed. As Watson wrote in that year's Annual Report, "In a very real

Charlie and Marie Robertson in the hallway of their house, now named Robertson House, in the late 1960s.

way, the creation of this Fund ranks as one of the most important events in our Lab's history." This was not the only gift Robertson made to the Laboratory. As we shall see in the next chapter, Robertson gave CSHL his Lloyd Harbor estate.

On Robertson's death in 1981, Watson wrote in tribute to him:

> In all his dealings with me and other members of the Laboratory, Charles S. Robertson was the perfect benefactor, modest, yet highly intelligent and desirous of further learning, gracious serious, but always with a sense of humor, and deeply loving and loyal to those individuals and institutions that he admired. His coming into our lives was a marvelous, unexpected gift, his delight and interest in the programs he helped make possible was a joy, and his passing this spring an occasion of deep sadness and reverence.

E.O. Wilson wrote in his 1994 autobiography that he "… wouldn't put him [Watson] in charge of a lemonade stand." Watson confounded such expectations, and by any measure, the first years of Watson's directorship were extraordinarily successful. From 1968 to 1975, the Laboratory's income rose from $640,000 to $3,778,000 and the number of grants from 11 to 35. With funding from federal grants and foundations and the stability provided by the Robertson Research Fund, the financial state of the Laboratory was strong. Just as importantly, the Laboratory's remarkable staff of young scientists made the future look bright indeed.

ETHIDIUM BROMIDE AGAROSE GEL ELECTROPHORESIS FUELS THE HUNT FOR MORE RESTRICTION ENDONUCLEASES

Summary

The innovation of using ethidium bromide in the early 1970s to visualize nucleic acids on agarose gels provided a faster, more sensitive, and more flexible method for viewing results and eliminated the time-consuming autoradiography. This was particularly important for identifying and using restriction endonucleases to map bacterial and viral genomes.

The Discovery/The Research

"Restriction modification" was the name given to the phenomenon, discovered in the 1950s that bacteria had endonucleases that cut and destroyed the DNA of infecting phage. The bacterial DNA was modified by the methylation to protect it so the action of these nucleases was restricted. The first type I restriction endonucleases to be isolated cut DNA at varying distances from the recognition sequence. The type II enzymes, first studied by Ham Smith, cut at the same nucleotide within the recognition sequence and thus the DNA was cut into discrete fragments. In 1971, Danna and Nathans showed that the enzyme *Hind*II cut SV40 DNA into 11 fragments (A–K) and commented that it should be possible to assign genes to these fragments and thus produce a map of the SV40 genome. The following year, they were able to assign the SV40 origin of replication to fragment C. Restriction mapping was thus taken up in earnest and a search began for restriction enzymes with different sequence specificities.

Isolating and purifying these new enzymes, however, was laborious and time-consuming. The activities of enzymes during purification were monitored by cutting radioactively labeled DNA, electrophoresis in polyacrylamide gels, and detecting the DNA fragments either by autoradiography or by slicing the gels and measuring the activity in each slice by scintillation counting. Alternatively, the DNA fragments were separated by velocity centrifugation on sucrose gradients. Sharp et al. set out to isolate restriction enzymes from *Haemophilus parainfluenzae* and brought together two existing techniques that eased the burden of detecting the DNA fragments produced by restriction enzymes. The first technique was the use of agarose gels rather than polyacrylamide gels and the second was the use of ethidium bromide to stain the DNA fragments by incorporating the dye in the electrophoresis buffer. Sharp et al. listed the advantages: Agarose gels can cope with buffers used to elute chromatography columns; very small quantities of DNA can be detected; the technique was rapid and cheap; DNA fragments were well resolved and their sizes could be determined with some accuracy. They concluded that "In our hands, the technique has proved to be a very useful and flexible tool for assaying restriction enzymes."

Significance

Recombinant DNA pioneer Herbert Boyer tells of a visit to the Sambrook and Sharp lab darkroom at Cold Spring Harbor where they showed him an agarose gel run with ethidium bromide visualizing cleaved adenovirus DNA. "It was one of the most exciting things I could have looked at," Boyer said. Boyer was not alone in recognizing the usefulness of this simple technique and it was rapidly adopted by molecular biologists the world over and is still used today.

The Scientists

Phillip A. Sharp was born June 6, 1944, in rural Kentucky. Sharp earned his B.A. in Chemistry and Mathematics at

Continued

Union College in Barbourville, Kentucky, and his Ph.D. in physical chemistry in 1969 at the University of Illinois. He was a postdoc with Norman Davidson at the California Institute of Technology and was at Cold Spring Harbor Laboratory from 1971 to 1974. Sharp moved to the Center for Cancer Research at the Massachusetts Institute of Technology in 1974. His many awards include the 1988 Albert Lasker Basic Medical Research Award and the 1993 Nobel Prize in Physiology or Medicine for his discovery of RNA splicing.

Joe Sambrook was born March 1, 1939, in Liverpool, England. After earning his undergraduate degree at Liverpool University, he went to Australian National University where he earned his Ph.D. in 1965 in the John Curtin School of Medical Research (incidentally, the institute where John Cairns had worked and where Bruce Stillman did his Ph.D.). After working at the MRC Laboratory of Molecular Biology and the Salk Institute for Biological Studies, he moved to Cold Spring Harbor Laboratory in 1969. Sambrook was Assistant Director from 1977 to 1985, when he left for Southwestern Medical Center, Dallas in 1985, returning to Australia in 1995.

Bill Sugden came to Cold Spring Harbor Laboratory in 1969 to work in Joe Sambrook's laboratory. His Columbia University Ph.D. was awarded in 1973 for the research he carried out at Cold Spring Harbor. Following his stay at Cold Spring Harbor, he went to George Klein's laboratory

Bill Sugden and Brad Ozanne relax on the Blackford lawn. Sugden and Ozanne were graduate students at CSHL in the early 1970s. Sugden received his Ph.D. in 1973, the first occasion on which research at CSHL had led to a Ph.D.

at the Karolinska Institute in Stockholm for a postdoctoral fellowship. Sugden returned to the United States to join the McArdle Laboratory for Cancer Research at the University of Wisconsin, Madison, where he has been Professor of Oncology since 1985.

— J.C.

AN AMAZING DISCOVERY REVEALS THAT GENES CONSIST OF PIECES—INTRONS AND EXONS

Summary

Scientists at Cold Spring Harbor Laboratory and the Massachusetts Institute of Technology (MIT) discovered RNA splicing, the construction of messenger RNA by cutting out the intervening sequences (introns) and joining just the coding parts (exons) of nuclear DNA in eukaryotes. Although genes were previously thought to be contiguous stretches of coding DNA that were collinear with its mRNA transcript (still largely true for prokaryotes), these experiments showed that eukaryotic genes were in pieces.

The Discovery/The Research

Several groups at CSHL were studying gene expression in adenovirus and getting results that were difficult to explain. For example, Ray Gesteland's group was isolating mRNAs and determining what they coded for by using the messenger RNAs to synthesize the proteins in the test tube. They selected mRNAs using pieces of DNA from known locations in the adenovirus genome and were surprised to find that DNA fragments from wide apart in the genome selected the same mRNAs.

Richard Gelinas, who was a postdoc in Rich Roberts' lab, had started to identify mRNAs that acted as the initiation and termination signals for the mRNAs. They made a "startling finding" that all late mRNAs began with the same oligonucleotide that was "not encoded on the DNA next to the main body of the mRNA." They concluded that either all RNA messages began with the same sequence or perhaps the experiment was done wrong. But repeated experiments gave the same result.

Louise Chow and Tom Broker, who arrived at CSHL in 1975, were also getting strange results. They were mapping mRNAs to the adenovirus genome by using electron microscopy to see where an RNA hybridized to the genome. They were surprised to find that many of the RNAs had 5′ ends that did not bind to the DNA, implying that those sequences were not part of the coding sequence.

All these anomalies were resolved by a collaborative experiment by Roberts, Gelinas, Chow, and Broker. The sequences of the free 5′ mRNA ends were identified by adding single-stranded restriction fragments from known locations in the genome. The results showed that the 5′ ends of the mRNAs were "complementary to sequences within the Ad2 genome which are remote from the DNA from which the main coding sequence of each mRNA is transcribed." As a consequence, the data results were "… not directly consistent with any mechanism previously suggested for the biosynthesis of mRNA in eukaryotic cells."

Meanwhile, at MIT, Susan Berget and Claire Moore in Phil Sharp's lab were performing almost identical experiments using electron microscopy and R-looping to map the hexon protein mRNA to the adenoviral genome. They found three loops of single-stranded DNA formed at the 5′ end of the hybrid molecule, indicating that this part of the mRNA was transcribed from three discontinuous positions.

The results from CSHL and MIT were presented in June at the 1977 Cold Spring Harbor Symposium on Quantitative Biology. A meeting report was immediately sent by Joe Sambrook to the journal *Nature* describing how the "audience at the Symposium was amazed, fascinated and not a little bewildered to learn that late adenovirus mRNAs are mosaic molecules consisting of sequences complementary to several non-contiguous segments of the viral genome."

In September four papers reporting the CSHL research appeared back-to-back in the September 1977 issue of *Cell*

Continued

while the MIT research was published in the August issue of *Proceedings of the National Academy of Science*. In it, Sharp coined the term "RNA splicing" and later recalled, "I can remember getting out my dictionary to see if this was going to be an appropriate term."

With these surprising results now out in the open, other investigators recognized splicing in other systems, including the chick ovalbumin gene, *Drosophila* ribosomal genes, globin genes, SV40, and yeast tRNA. In 1978, Walter Gilbert coined the terms "intron" for intervening sequence and "exon" for the coding regions, "exported" to the cytoplasm.

Significance

Ever since the realization that triplets of nucleotides coded for amino acids, it had been assumed that the order of the triplets in a gene was collinear with the order of amino acids in the corresponding protein. As Brenner et al. put it after the discovery of messenger RNA, the latter was a simple copy of the gene. The finding that genes were made up of coding and noncoding sequences was revolutionary and called for a radical revision of what constitutes a gene. It also opened up new areas of research into the mechanism of splicing, alternative splicing, errors in splicing cause disorders, and not least evolutionary implications: Why was there noncoding DNA?

The Scientists

Richard J. Roberts was a Senior Staff Investigator at CSHL from 1972 to 1986, and Assistant Director for Research from 1986 to 1992, and then went to New England BioLabs. He has had numerous awards and honorary degrees, including a knighthood in 2008, 1997 Fellow of the American Society of Arts and Science, and 1995 Fellow of the Royal Society.

Louise Chow and Tom Broker came to CSHL in 1975, Tom as Chief of the Electron Microscopy Section and

Richard Gelinas in 1978.

Louise as Staff Investigator; they were later appointed Senior Staff Investigators. They left in 1984, first to Rochester School of Medicine and then the University of Alabama at Birmingham, where they study human papillomaviruses.

Richard Gelinas obtained his Ph.D. in Fotis Kafatos's laboratory at Harvard in 1974. That year he moved to Cold Spring Harbor Laboratory as an NCI Postdoctoral Fellow to Rich Robert's laboratory to characterize the sequences initiating and terminating adenovirus gene transcription, He was promoted to Staff Scientist before leaving in 1979 for the Fred Hutchinson Cancer Research Center in Seattle. Gelinas spent some time in industry at ICOS Corporation and Celltech R&D. Since 2009 he has been at the Institute for Systems Biology in Seattle.

— *J.C.*

15

A Period of Expansion, 1978–1988

There were significant changes at the Laboratory as it entered its second decade under Watson's leadership. There was an expansion of research fueled in part by the increasingly sophisticated genetic manipulations made possible by recombinant DNA techniques. Restrictions on gene cloning had been relaxed by the late 1970s, and, by the early 1980s, cloning and analyzing genes from eukaryotic cells, whether yeast or human, had become, if not routine, at least commonplace. This was due, at least in part, to laboratories the world over, guided by the best-selling CSHL manual *Molecular Cloning,* by Maniatis, Fritsch, and Sambrook, doing experiments that had previously been restricted to a few centers. The new tools of molecular biology—DNA sequencing and synthesis, protein sequencing, monoclonal antibodies, and transgenic mice—all brought a level of experimental sophistication undreamt of in the previous decade.

CSHL continued to be at the cutting edge of research, and by 1981 new areas of research included yeast genetics, mammalian genetics, plant science, and neurobiology. Research on tumor viruses continued, but it was less concerned about what the viruses might reveal about cancer than using them as experimental organisms in their own right for studying fundamental biological processes like DNA synthesis and gene expression. New groups proliferated: There were eight independent laboratories in 1978; by 1988 there were 28. These changes in research were reflected in changes in scientific staff, as the number of investigators, postdocs, and graduate students rose from 56 in 1978 to 107 in 1988. The increasing numbers hide the considerable turnover that occurred in this

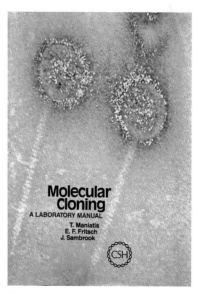

The cover of first edition of Molecular Cloning.

259

period. Of the 24 investigators at CSHL in 1978, only four (Grodzicker, Mathews, Roberts, and Wigler) remained in 1988.

The most significant departure was that of Joe Sambrook who had been recruited by Watson in 1969 and had established the Tumor Virus Group in James Laboratory as a powerhouse of CSHL and cancer research. Sambrook had taken over from Ray Gesteland as Assistant Director in 1978 and was acting director when Watson took a sabbatical year in London in 1983. By all accounts their relationship had at times been tempestuous. Perhaps Watson felt Sambrook had overstepped his authority when Sambrook made changes to the *1983 Annual Report*, the more heinous being printing it in a different size and printing the reports of meetings on colored paper. Sambrook went first to the University of Texas Southwestern Medical School as Chairman of biochemistry, and then he returned to Australia in 1995 as Director of Research at the Peter McCallum Cancer Centre in Melbourne.

Sambrook came back to CSHL in the summer of 1985 when the new extension to James Laboratory was named in his honor. The occasion

The dedication ceremony for the Sambrook Laboratory addition to James Laboratory, 1985.

achieved much wider notice because of the speech given by Renato Dulbecco and printed in *Science* in March 1986. Reflecting on the state of cancer research, he argued that there were two ways to find cancer genes: doing it piecemeal or by sequencing the whole genome of a selected animal species. And, if we want to understand human cancer, that species must be *Homo sapiens*. It was from this small beginning that the Human Genome Project grew, with profound consequences for the world and for Cold Spring Harbor.

Barbara McClintock Receives the 1983 Nobel Prize in Physiology or Medicine

McClintock had received increasing recognition for her work through the 1970s—in 1971, the President's National Medal of Science and honorary doctorates from Harvard and Rockefeller Universities—but the honors became a flood in the early 1980s. In 1981 she received the Thomas Hunt Morgan Medal, the Wolf Prize in Medicine, and the Albert Lasker Basic Medical Research Award and was named a MacArthur Prize Fellow Laureate. In 1982, she received the Louisa Gross Horwitz Prize for Biology or Biochemistry and the Charles Leopold Mayer Prize. Her honors culminated in 1983 when she received the Nobel Prize for Physiology or Medicine, joining Marie Curie and Dorothy Hodgkin as the only women to receive unshared Nobel prizes.

Charles Robertson Gives His Estate to the Laboratory

Now that the research programs were on a firmer footing, it is not surprising that Watson should turn his attention to the educational work of the Laboratory. As long ago as 1969, he had argued that the Laboratory could maintain its position as the place to come for exchange of information only by greatly expanding the number of meetings and courses. It was Charles Robertson who provided the means to do so.

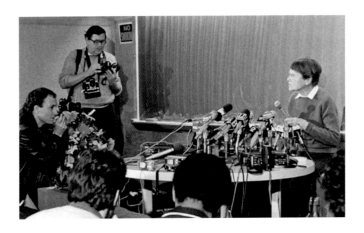

Barbara McClintock at the press conference in Bush Auditorium following the announcement of her Nobel prize, October 12, 1983.

In 1975 Robertson proposed donating his Lloyd Harbor estate and its buildings to the Laboratory so that a Center for Neurobiology could be created there. Watson persuaded him that the Village was not likely to sanction the construction of laboratories within its boundaries and that the costs were too great. Instead Watson, drawing on his experiences as a participant in the Ciba Foundation Symposia in London, proposed that the estate be used as a conference center for meetings of 24 to 36 scientists, thus complementing the 10-fold larger meetings on the main campus.

In June 1975, the Laboratory's trustees passed a motion "… enthusiastically and with gratitude" accepting Robertson's gift. Robertson also thought of the future. Such a gift, although munificent, can be a burden on the recipient institution that has to find funds for its upkeep. So in addition to the property, he gave an endowment of $1.5 million, the income from which was to be used for maintenance of the property and to pay the equivalent of village taxes. Work began converting the garage to a conference room and was completed in 1976 in time for Francis Crick, who was attending the Symposium on Chromatin, to open the Center with a talk on "How Scientists Work," emphasizing the importance of scientists talking to one another and the need for facilities like the Banbury Center.

The Robertson estate garage being converted into the Banbury Conference Room, 1976.

Watson decided that the program for the Center should focus on car-cinogenesis, environmental hazards, and risk assessment, a decision based on the contemporary interest in those topics. In 1978, Victor McElheny, a science reporter who was especially interested in environmental issues, became the first director of the Banbury Center. The first meeting was on "Assessing Chemical Mutagens: The Risk to Humans" and was held on May 15–17, 1978. McElheny also instituted a series of workshops for science journalists and Congressional staff, funded by the Alfred P. Sloan Foundation. Representatives from each group came to Banbury for two days to learn in depth about some scientific issue of public interest.

Although the emphasis in the first years was on environmental issues, the range of topics expanded when Michael Shodell became director in 1982. Shodell returned to university teaching in 1986, and his place was taken in April 1986 by Steve Prentis who also took on the Laboratory's publishing program (see page 267). Tragically, Prentis was killed in a car crash one year later, and I took his place, coming from the Institute of Molecular Genetics at Baylor College of Medicine. It was an exciting time with the beginnings of the Human Genome Project and

Eric Lander, Rich Roberts, and Peter Neufeld examining a Southern blot, DNA Technology and Forensic Science meeting, Banbury Center, 1988.

early successes in locating and cloning human disease genes. Banbury became known for meetings on topics like the genetics of Alzheimer's disease, schizophrenia, manic depression, breast cancer, and muscular dystrophy.

Some of the most interesting meetings were those on topics at the intersection of modern genetics and society. The most important such meeting was DNA Technology and Forensics, held in 1988. It was the first time that DNA fingerprinting by forensic scientists was subjected to critical scrutiny by molecular geneticists, and it led to significant changes in the way the DNA fingerprinting was implemented. Moreover, Barry Scheck and Peter Neufeld were participants and were encouraged to found the Innocence Project, which uses DNA data to free the wrongly convicted.

The DNA Learning Center—Genetics Education for High School Students

The Banbury Center was not the only new educational initiative in this period. While its meetings are intended for the highest level of professional scientists, Dave Micklos began the Laboratory's DNA Literacy Program for high school students.

With a background in science writing, Micklos had been hired by Watson in 1982 to set up a department of Public Affairs and Development. In 1984, Micklos took a phone call from Fran Roberts, the superintendent of Cold Spring Harbor School District, suggesting that the school district and the Laboratory might join forces to improve local science education. An application to the NSF for funding to train teachers to do recombinant DNA experiments failed, but eight local schools provided funds to purchase the necessary equipment and consumables. Micklos and Greg Freyer, a postdoc in Rich Roberts' laboratory, devised two simple experiments that required relatively little equipment: cutting DNA with restriction enzymes and running the fragments on a gel, and inserting an antibiotic resistance gene into a plasmid. Micklos and Freyer taught their first DNA classes in the summer of 1985. The "Vector Van" appeared the following year, crammed full of equipment, and was used to conduct weeklong teacher workshops at six sites across the nation.

Dave Micklos, summer intern Jeff Diamond, and Greg Freyer outside Jones Laboratory with the two Vector Vans, 1987.

Watson, recognizing the uniqueness of this project and its considerable value in promoting CSHL to a new constituency, decided that it needed a building of its own. Fortuitously, there was a suitable building, the former Cold Spring Harbor School on Main Street in Cold Spring Harbor, just two miles from the Laboratory. Renovations began in 1987, and the following spring the first students came on a day trip to the DNA Learning Center (DNALC) to do experiments. The DNALC flourished, and in 2001 an extension was added to the building, providing extra teaching laboratories and a much needed eating area. The Center became the Dolan DNA Learning Center in honor of the support given by Charles and Helen Dolan.

A key feature of the DNALC program is the extensive use made of web-based projects. The first was "DNA from the Beginning," and there are now more than 20 websites, including one on the American eugenics movement. In 2014, more than 6.7 million visitors used the DNALC multimedia materials.

The program has proven tremendously successful; some 265,000 precollege students have conducted experiments in the Dolan DNALC, and a further 150,000 have received instruction in their own schools. The outreach is not restricted to students near Cold Spring Harbor. The DNALC has been the direct model for centers in South Carolina, Indiana, and California, as well as in Australia, France, Italy, Singapore, Germany, Austria, England, and China.

The Laboratory's Publishing Expands

As early as 1975, Watson had written that the Laboratory ran a "mini-university press type operation," at that time publishing three or four books each year in addition to the Symposium volume. Moreover, he pointed out, these were substantial books, rarely less than 600 pages long and often more than 900 pages long. Watson himself initiated most of them, seeing the burgeoning number of topic-based conferences at the Laboratory as an opportunity to press distinguished members of each research community into service as editors of a valuable, often unique reference work for their field. By 1985, the publishing group was rather more than a mini-press, publishing 11 new books, five new editions, seven reprints, and 10 abstract books for meetings and dealing with marketing, warehousing and shipping, and foreign rights.

This expansion was fueled in part by a new type of publication in biology, the laboratory manual, which provided detailed, step-by-step instructions on methods and protocols. These manuals were products of the Laboratory's growing array of practical courses and the first, *Experiments in Molecular Genetics*, by Jeffrey Miller, published in 1972, was based on the Bacterial Genetics course. The most significant manual was the best-selling *Molecular Cloning* by Tom Maniatis, Ed Fritsch, and Joe Sambrook. First published in 1982 and derived from the Molecular Cloning of Eukaryotic Genes course, it introduced the larger community of scientists to what had been esoteric techniques for manipulating DNA. Known as the "bible" for its authority, the first edition sold more than 60,000 copies in 7 years and promoted the use of recombinant DNA worldwide. Subsequent editions have brought total sales to more than 200,000 copies.

The operations of the publications department had been managed by Nancy Ford since 1973, but as the Laboratory grew Watson had less time to initiate new publishing projects and recognized the need for additional help. In 1985, Steve Prentis came to the Laboratory as director of the Banbury Center. He was an experienced editor of Elsevier's monthly *Trends* review journals and brought with him the idea of establishing a journal at Cold Spring Harbor in collaboration with the Genetics Society of Great Britain. He took on responsibility for both the publishing department and Banbury and oversaw the launch of the journal *Genes & Development* in March 1987. Tragically, Prentis never saw a copy of the journal—he was killed in a car accident one week before its first appearance. Staff scientist Mike Mathews stepped in as the U.S. editor with Grahame Bulfield as European Editor, and they guided the journal through its early years until staff scientist Terri Grodzicker became U.S. editor, and later sole editor, in 1989.

Watson realized that his vision of growth for both the publishing operation and the Banbury Center required separate leadership, so when I came to Banbury in 1987, John Inglis came to take over publishing. Inglis had been an Assistant Editor of *The Lancet*, the founding editor of *Immunology Today*, and the initiator of several other Elsevier journals. One of his first acts as Executive Director was to give publishing its own distinctive identity, the "Cold Spring Harbor Laboratory Press," which was founded in January 1988.

His goals for the Press included commissioning new monographs and manuals, ensuring new editions for the strongest titles, and improving production efficiency, marketing effectiveness, and financial success. The rapid rise in the scientific importance and financial success of *Genes & Development* encouraged the launch of other journals and, by 1995, *G&D* was joined by *Genome Research* and *Learning & Memory*. The Press also published journals on behalf of two societies—The Protein Society between January 2001 and December 2008 and since January 2003, The RNA Society.

Jan Witkowski and John Inglis, 1987.

For many years, publishing's editorial offices had been based in Urey Cottage, but with the expansion of Press projects and additional staff, it became increasingly crowded, while other parts of the Press operation were scattered on and off the campus. In 2001, the Laboratory moved into the former American Institute of Physics building in Woodbury. Part of the building became the Genome Center, and one wing was used to house nearly all the Press staff efficiently under one roof.

Since the advent of the web browser 20 years ago, publishing has changed immeasurably, and CSHL Press has undergone its own digital transformation. It was the first science publisher to have a website, in May of 1994. It spun out a venture-capital-backed internet startup, BioSupplyNet.com in 1995. *G&D* and *Genome Research*, long ranked among the top research journals in the world, and the other journals have been published online since 1997, and a new open-access clinical journal *Molecular Case Studies* will be published online-only in 2015. The third edition of *Molecular Cloning* had a website in 2000 that became the foundation of a successful new journal *Cold Spring Harbor Protocols*. The much-valued Cold Spring Harbor monograph series inspired two *Perspectives* journals in biology and medicine with an innovative online-plus-print-book publishing model. The Press has also made an important contribution to open science with the launch in 2014 of bioRχiv, the first centralized preprint service for biology.

The Laboratory's publishing activities began in 1934 with the first of the Symposium volumes. The intellectual and financial benefits these brought to the Laboratory have increased many-fold as the Press continues to provide books, journals, and services that help scientists succeed in their research. Together with the Meetings and Courses programs, the Press enhances the Laboratory's reputation as an internationally renowned center for information in the biological sciences

The Symposium Remains the Flagship Meeting

The early Symposia of the Watson years tended to focus on molecular biology (e.g., *The Mechanism of Protein Synthesis* and *Transcription of Genetic Material*) and topics relevant to the Laboratory's research program (e.g., *Tumor Viruses* and *The Synapse*). However, increasingly the topics moved from research in molecular biology to molecular studies of biological function. Examples include *Origins of Lymphocyte Diversity*, *Molecular Biology of Development*, and, in 1986, *Molecular Biology of* Homo sapiens. The latter was notable for the discussion of the Human Genome Project (HGP) in the newly opened Grace Auditorium. The

Wally Gilbert at the chalkboard in Grace Auditorium during the discussion of the Human Genome Project at the 1986 Symposium The Molecular Biology of Homo sapiens.

Pre-banquet cocktails at the 1986 Symposium on Molecular Biology of Homo sapiens. *Kary Mullis is in the center, gesturing.*

first discussions about a human genome project had been held in a number of small, closed meetings, and the discussion in Grace Auditorium was the first open to rank-and-file scientists. Watson inserted the discussion just before the banquet cocktails and it proved a lively occasion. When Wally Gilbert estimated that the project might cost $3 billion, the audience was stunned, believing that the only source of this money would be the R01 grants supporting their research. There was a lighter moment when Kary Mullis pointed out that G and C were difficult to distinguish in computer printouts of DNA sequence and made the suggestion, regrettably not adopted, that as Crick already had a nucleotide (C), that G should be replaced by W for Watson. Mullis's more serious presentation was on the polymerase chain reaction, the first time that many in the audience became aware of its power.

The nature of the Symposia also changed in this period. They had been a primary source of information—there were few journals, and if you wanted to know the latest in molecular biology, the Symposium was the place to be. Then the world of molecular biology was still small enough that most of the leaders could and did attend the Symposia. But through the 1970s and 1980s, as the number of scientists using molecular techniques rose rapidly, so did the number of journals. A newcomer, *Cell*, founded by Ben Lewin in 1972 became the hot journal. The Symposia gradually became reviews of a field, becoming larger and larger until the proceedings had to be published in two volumes.

Regrettably, at least to those of us who fondly remember the library shelves bending under the weight of a long line of dusky red volumes, the printed volumes may become a thing of the past in a world of online publishing.

The Meetings and Courses Programs Continue to Enlarge

For many years the Symposium was the only meeting held at Cold Spring Harbor until, in 1950 and 1951, Delbrück organized a two-day meeting on Phage, following the end of the Phage course. Curiously, while the Phage meetings were held throughout the 1950s and 1960s, they seem not to have been regarded as "official" meetings. They were listed briefly under the title of "Professional Meetings," and the abstracts were published in the Phage Information Service bulletins.

Watson began enlarging the meetings program with the addition of meetings in 1969, for a total of four that year. One was on the Lactose Operon, which had played a key role in molecular approaches to gene expression, for which Jacques Monod and François Jacob together with André Lwoff had won the 1965 Nobel Prize in Physiology or Medicine. The second meeting was on Tumor Viruses, a meeting that under various names has been held annually ever since. The latter soon

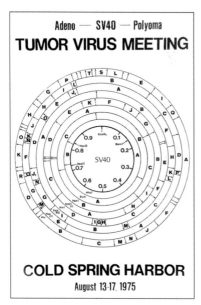

A restriction map of SV40, the RNA Tumor Virus Meeting poster, August 13–17, 1975.

became so popular that it was split into DNA Tumor Viruses and RNA Tumor Viruses. It was an example followed in subsequent years when the topic of an oversubscribed meeting would be split and spawn new meetings. By 1977, the number of meetings had reached double figures and a pattern had been set of a small number of annual and biannual meetings, with other meetings held only once. The meeting topics did not necessarily reflect the research interests of the Laboratory. For example, the 1980 program included meetings on hematopoietic cells, herpes virus, mitochondrial genetics, and muscle development, none of which were topics of research at CSHL. One of the most influential of the new meetings was the Genome Mapping and Sequencing meeting, which began in 1988. This meeting rapidly became the not-to-be-missed meeting in the genomics world and contributed significantly to the development and growth of the Human Genome Project.

There had been no major change in the summer courses since 1958, when Demerec added the mammalian cell culture course, but in 1969, the number of courses had increased to five, by 1979 to 14, and by 1989 to 19. The large increase in numbers in this period reflected the introduction of recombinant DNA techniques and modern refinements of older techniques. The most significant changes in topics between 1969 and 1979 were the elimination of the viral and bacterial genetics courses—only the Advanced Bacterial Genetics course survived—and the addition of six neuroscience courses. Other new courses taught techniques common to many areas of biology: Macromolecular Crystallography, Advanced Molecular Cloning, and Protein Purification and Characterization. There were two courses on human diseases; one, with Eric Lander and David Page as instructors, introduced students to the latest in molecular mapping of human genetic disorders.

For 20 years Watson oversaw the meetings program, but in 1986 Terri Grodzicker took over responsibility for meetings and courses when she was appointed Assistant Director for Academic Affairs.

The cover of the abstract book for the 1991 Genome meeting. Salesman Watson pitches the Human Genome Project to Leroy Hood, David Botstein, Sydney Brenner, and Charles Cantor.

Cold Spring Harbor Laboratory Fellows

One of the key steps in the career of a young scientist is the move from postdoctoral fellow to an independent researcher. This requires acquiring the funds needed to finance the research, funds that usually come from NIH. However, in the wake of the Gramm–Rudman–Hollings Balanced Budget Act of 1985, Watson worried that this support would be increasingly difficult to get so that same year the Laboratory established the Cold Spring Harbor Laboratory Fellows Program to provide exceptional postdocs with three years' support for independent research. During this time they would publish papers on their research, thus having a strong basis for applying for NIH grants.

The program has been extremely successful. The first four fellows were Adrian Krainer (1987), Carol Greider (1988), Eric Richards (1989, and David Barford (1991). Krainer, who studied RNA splicing, became an investigator at CSHL, and developed a therapy for spinal muscular atrophy (Chapter 17). Greider came from Elizabeth Blackburn's laboratory at the University of California, San Francisco, to continue working on telomere biology and the relationship between telomeres and aging. She and Blackburn shared the 2009 Nobel Prize in Physiology or Medicine with Jack Szostak. Richards worked on plant epigenetics and became Vice President for Research at the Boyce Thompson Institute. Barford used X-ray crystallography to determine the structure of the enzyme tyrosine phosphatase being studied in Nick Tonks' laboratory. He went to the Laboratory for Molecular Biology in Cambridge and was appointed a Fellow of the Royal Society in 2006.

Yeast Brings Eukaryotic Molecular Genetics
to Cold Spring Harbor

The first significant move to studying eukaryotic cells came with the establishment in 1978 of a group working on yeast. Just as Watson used tumor viruses to probe the workings of the cancer cell, so he had advocated using yeast as a first step to understanding more complex eukaryotic cells. In the second (1970) edition of *Molecular Biology of the Gene,* he wrote: "There are now so many reasons to intensify work on organisms like yeast … even if our primary interest is the human cell, this may be the time for many more biologists to work with organisms like yeasts." In that same year he invited Gerry Fink, Fred Sherman, and Bruce Lukins to teach a course on the "Molecular Biology and Genetics of Yeast." (Fink and Sherman taught the course for 17 years, and the course will be celebrating its 45th anniversary in 2015.)

Watson recalls that Ira Herskowitz phoned him to tell him that two of Herskowitz's students had obtained exciting results about mating-type switching from α to **a**, the equivalent of changing sex in yeast. Jim Hicks and Jeff Strathern had an idea about how this happened, and Herskowitz thought that they and CSHL would benefit from their working at Cold Spring Harbor. Hicks and Strathern moved from Oregon and were joined by Amar Klar, who had been working on the same problem at Berkeley.

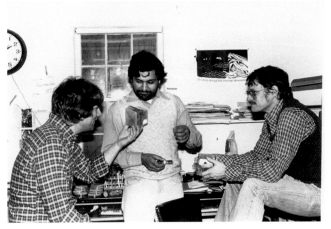

Jim Hicks, Amar Klar, and Jeff Strathern, 1980.

Jim Broach, who was working on yeast galactose genes in Ray Gesteland's group, was the fourth member of team, although he left in 1979 for SUNY Stony Brook. They set up shop in Davenport Laboratory in April 1978 but had to move out almost immediately to make room for the summer courses, which they had to do every year until 1981 when they took up residence in Delbrück Laboratory, a new extension added to Davenport Laboratory. Despite the inconveniences of the old Davenport Laboratory, they were able to show their idea (the "cassette" model) was correct—yeast change mating type by moving around pieces of DNA, just like the movable elements studied by Barbara McClintock.

The appointment of David Beach in 1982 ensured that yeast remained an important experimental organism at CSHL after the founders of the yeast group left—Strathern in 1983, Hicks in 1985, and Klar in 1987.

David Beach, circa 1989.

David Beach had collaborated with Paul Nurse at the University of Sussex, showing that the *cdc2* and *CDC28* genes that control the cell cycle of the fission yeast *Schizosaccharomyces pombe* and the budding yeast *Saccharomyces cerevisiae*, respectively, are homologous. At CSHL, Beach continued to dissect the genes and proteins controlling the yeast cell cycle, research that occupied him until he left CSHL in 2000. In 1988 Beach and his collaborators showed that MPF, a protein discovered in the toad *Xenopus laevis* that was required for a cell to enter mitosis, contained the toad equivalent of the yeast cdc2 protein. Later, Beach's laboratory made another important connection when they showed that cdc2 associated with the cyclins, proteins that oscillate during the cell cycle.

In 1981, there was a small but significant change in the name of the Yeast group. "Plant" was added to the title when Stephen Dellaporta and Russ Malmberg joined the group. Plant cells now could be grown in tissue culture and genetically engineered using the Ti plasmid *Agrobacterium* system, so that plant cells could now be genetically manipulated in the same ways as yeast or bacteria. A significant shift in research began the following year when Dellaporta and Hicks turned their attention to a molecular analysis of McClintock's discovery of transposition in maize. This became a long-running research project of the plant group, taken up by Venkatesan Sundaresan when Hicks and Dellaporta left in 1986. Three years later, Sundaresan was joined by Rob Martienssen. Together they went on to develop an enhancer and gene trap system in *Arabidopsis* using McClintock's transposable elements and employed the system to identify developmentally expressed genes.

For many years the research of the plant scientists was subsumed under "Molecular Genetics" or "Genetics," and it was not until after a major expansion during the 1990s that in 2001 it was recognized as an independent division.

Tumor Viruses Continue to Be a Major Focus of Research

The discovery by Harold E. Varmus and J. Michael Bishop in 1979 that Rous sarcoma virus had a gene, *v-src,* that was a mutated version of the cellular gene for a tyrosine kinase, led to a model in which it was proposed that cancer resulted from abnormalities of cellular genes controlling key functions like cell division. This led to a search for other examples of cellular genes hijacked by cancer-causing RNA viruses. Nevertheless DNA tumor viruses still had a contribution to make to cancer research. There were "native" genes of SV40 and adenovirus encoding proteins (oncoproteins) that made normal cells behave like cancer cells, and research on these proteins became the focus of the Tumor Virus Group in the late 1970s and 1980s.

Robert Tjian from Harvard joined the Tumor Virus Group in 1976 and remained through 1978. His goal was to determine the function T antigen, a protein made by SV40 and necessary to convert a normal cell to a cancer cell. Tjian used an adenovirus–SV40 hybrid virus to produce large amounts of a hybrid protein that he could use in functional assays. Using DNA protection assays, Tjian showed that T antigen was a DNA-binding regulatory protein. Subsequently the functions of the T antigen in infected cells were studied, and it was shown to bind to key regulatory cellular proteins including p53 and Rb, the protein products of two tumor-suppressor genes.

Robert Tjian, ca. 1978.

T antigen subsequently played a role in two other projects at Cold Spring Harbor that were important for cancer research. Yakov ("Yasha") Gluzman came to work in James Laboratory in 1977. In 1980, he developed a line of monkey kidney cells into which he had introduced an SV40 mutant with a mutation at the origin of replication. This meant that although the virus could not replicate, the cells could produce T antigen. If the cells were transfected with, for example, a human gene engineered to carry the SV40 origin of replication, the gene would be replicated and

Yasha Gluzman in his element at the 1983 Enhancers and Controlling Elements meeting at the Banbury Center.

Doug Hanahan, 1981.

the human protein made in the cells. This proved to be a very timely development in biotechnology when it became clear that making mammalian proteins in bacterial cells was not as easy as had been expected. Yasha left CSHL in 1989 to work at Lederle Laboratories, Pearl River. Yasha's life was tragically short; in 1996 his wife persuaded her cousin to murder him.

Some years later, Doug Hanahan used T antigen in analogous fashion, but in mice rather than in cells in culture. Hanahan came to Cold Spring Harbor in 1979, moving to James Laboratory and the Tumor Virus Group in 1980. He was the first at CSHL to make transgenic mice (i.e., mice containing genes from other organisms). Hanahan was able to produce mice in which the SV40 T antigen gene was expressed only in the islet cells of the pancreas, the cells that make insulin. As a consequence, the mice developed pancreatic tumors. The work resulted in three major findings. First, it demonstrated that expression of a transgene could be directed to a specific cell type in vivo. Second, it showed that a cancer gene identified through cell culture experiments could produce tumors of cells in the animal. And, third, because not every transgenic pancreatic cell making T antigen gave rise to a tumor, it was clear that cancer was a multistep process. Hanahan's paper also highlighted the importance of tumor vasculature expansion to tumor growth.

The Tumor Virus section of the 1990 Annual Report was prefaced with the comment that "... the ground covered by this section has expanded considerably over the years," in part as SV40 and adenovirus were used to understand fundamental processes in biology. For example, Bruce Stillman who arrived as a postdoc in 1979, used SV40 to analyze the proteins involved in DNA replication. In 1987, he, his graduate student Greg Prelich, and colleagues Dan Marshak and Mike Mathews identified a protein essential for replication of SV40 DNA, a protein identical to an already known protein, PCNA. Stillman continued purifying the many components of the protein machine that acts at the DNA replication

Mike Mathews, with postdocs Fred Asselbergs and Bruce Stillman, 1980.

fork and in 1992, he and Steve Bell revealed a multisubunit protein that binds to DNA and initiates replication. Stillman also studied chromatin, developing the first DNA replication-complex in vitro system for assembly of chromatin. SV40 was also studied for insights into fundamentals of gene expression. Enhancers are mysterious DNA sequences that can increase the expression of genes far away from the enhancer. Winship Herr showed that the SV40 enhancer was not a single unit but was composed of three elements that could compensate for each other.

Ed Harlow had done his Ph.D. with Lionel Crawford at Imperial Cancer Research Fund (now Cancer Research UK) in London using monoclonal antibodies to identify proteins in transformed cells that interact with and make complexes with SV40 T antigen. Harlow continued using the same strategy at Cold Spring Harbor but extended the work to include adenovirus E1A protein and p53. By 1986, Harlow's laboratory, including graduate students Karen Buchkovich and Peter Whyte, had identified a protein with a molecular weight of 105 kDa that coprecipitated with E1A when cell extracts were treated with monoclonal antibodies against E1A. In 1987, Wen-Hwa Lee's laboratory reported their work on the retinoblastoma (Rb) gene product. (Rb is a tumor-suppressor gene—mutations inactivating Rb cause tumors.) Harlow, in London at the time, noticed that the Rb protein had a molecular weight of ~110 kDa. He returned to find that Buchkovich and Whyte had made the same connection and had already started experiments that would show that the 105-kDa protein bound by E1A was indeed the Rb protein. Thus a connection was made between oncogenes and tumor-suppressor genes—the E1A protein causes cancers by binding to and inactivating the Rb protein.

Ed Harlow at the 1994 Symposium.

Cancer and Mammalian Cell Genetics

One of the key techniques that enabled the application of recombinant DNA technology to mammalian cells was transfection, in which cells in culture take up foreign DNA. Mike Wigler, Saul Silverstein, and Richard Axel at Columbia University had first developed co-transfection, in which the gene for a selectable marker is added together with the gene of interest. Cells taking up the marker can be selected, and a large proportion of these also take up the second gene, which can become stably integrated into the host cell genome. The technique proved invaluable in biotechnology as well as academic laboratories.

Wigler brought these skills to Cold Spring Harbor in 1978 and by 1980 had begun transfecting cells with DNA from human tumors. The following year, Wigler's laboratory was able to report that DNA extracted from five independent human tumors could transform cells, and that they had isolated the transforming gene from the T24 bladder carcinoma cell line. (Robert Weinberg and Geoffrey Cooper had also shown that cells could be transformed by DNA isolated from chemically and virally infected cells.) The genes isolated from these human tumors were related to the *ras* oncogenes carried by two murine sarcoma viruses. The human genes were called *HRAS* and *KRAS*. In 1982, Wigler's laboratory isolated a third gene, *NRAS*.

It was a period of intense excitement and activity with several other laboratories in the chase to determine how these cancer-causing versions of the human *RAS* gene differed from the normal version. The laboratories of Weinberg, Mariano Barbacid, and Wigler all published more or less simultaneously that the difference lay in a mutation of a single nucleotide, leading to the change of a single amino acid, glycine, to valine. It was, an editorial in *Nature* reported, "one of the most startling discoveries so far in the long and frustrating search for an understanding of cancer."

However, much remained obscure. What were the functions of the protein? How could a mutant protein initiate such profound changes in a cell? Jim Feramisco, who came to CSHL in 1978 as a postdoctoral fellow in Keith Burridge's cell biochemistry group, took a direct approach. He had been studying the cellular distributions of cytoskeleton proteins

Mike Wigler and David Kurtz, circa 1980.

Jim Feramisco, 1985.

by microinjecting fluorescently labeled proteins such as α-actinin into living cells in culture. In 1983, Feramisco began microinjecting purified RAS proteins into cells that had stopped dividing because they were packed together on the surface of a tissue culture dish. He found that injection of the mutant H-*ras* protein caused these cells to start dividing and synthesize DNA, whereas injection of the normal protein had no effect.

A second strategy was to use an organism more tractable than vertebrate cells in tissue culture and Wigler turned to yeast. Scott Powers from Columbia University joined Wigler's laboratory, and Wigler sent him down Bungtown Road to learn yeast biology from Hicks and his colleagues. Powers' strategy was simple—use the human *RAS* genes as probes to detect the yeast *RAS* genes in clone libraries of yeast DNA. Powers found two yeast *RAS* genes and, remarkably, the first 80 codons of the yeast and human *RAS* genes were 90% identical. They were also functionally equivalent; the human *RAS* genes could rescue mutant yeast cells lacking their *RAS* genes. Considering that the last common ancestor of yeast and human beings existed 1.5 billion years ago, this degree of conservation is extraordinary. Thus, hopes were high that the down-stream and upstream pathways would also be highly conserved and that yeast cells could be used to explore the functions of RAS.

Subsequent research in Wigler's and many other laboratories showed that the ras protein is a key component in signal transduction pathways that transfer signals received at the cell surface to the nucleus, affecting the expression of genes controlling DNA replication and cell division.

Jim Garrels Refines a Method for Analyzing Thousands of Cell Proteins at a Time

The interlaboratory collaboration between Wigler's group and the Yeast group was characteristic of research at CSHL, especially for laboratories with techniques that were generally applicable. One example was Jim Garrels' work on two-dimensional gel electrophoresis (2DGE), a rather unusual project for CSHL in that it was devoted to developing a technique. 2DGE had been developed by Pat O'Farrell at the University of Colorado to analyze simultaneously all or many of the proteins of a cell. In the first step, proteins were separated according to their size, and in the second step, according to their electric charge. The technique provided a means to compare the proteins of two different cell types (normal vs. cancer) as well as the changes in the proteins synthesized over time.

Garrels had made significant improvements in 2DGE while at the Salk Institute and came to CSHL to further refine the technique while applying it to biological problems relevant to the interests of CSHL scientists. A laboratory was fitted out to Garrel's needs and equipped with purpose-built apparatus designed to handle the large slab gels needed for 2DGE. Garrel's group wrote software to compare the patterns of proteins on the gels, and the Laboratory acquired its first powerful computer, a PDP-11/60 with an 88-MB hard drive, to carry out the analyses.

Jim Garrels, Quest Computer Lab, 1978.

(In 1977, the basic PDP-11/60 came with 256K memory and a 28-MB disk and cost $44,700!) Garrels collaborated with Laboratory scientists, particularly Bob Franza, as he refined the technique, developing the QUEST system, an integrated system of hardware and software for 2D-gel electrophoresis.

Commercial Applications of CSHL Discoveries

Garrels' work was also significant as being the basis for the first spinoff company based on CSHL research. In 1981, Garrels and Franza raised funds from Bear Stearns to found Protein Databases Inc., which initially provided a service analyzing samples prepared and sent by clients.

This was not the first occasion on which CSHL scientists had considered a commercial venture. In 1974, Rich Roberts had suggested to

Watson that the Laboratory should set up a laboratory off site to produce and sell restriction enzymes, relieving the burden on his laboratory and providing funds for CSHL. Watson decided it was inappropriate for an academic institution to do so, a decision later regretted. Instead, Roberts became consultant and chair of the scientific advisory committee of a new start-up, New England BioLabs, founded by Don Comb.

By 1980, times had changed with the founding by academic scientists of companies like Genentech, and CSHL joined in discussions about being part of Celltech, a British company, but these came to naught. Instead, in 1980, the Laboratory established an entity called the Cell Biology Corporation (CBC) to act as an intermediary between the academic research at Cold Spring Harbor and companies. Tissue plasminogen activator (TPA), a blood clot "dissolver," was chosen as the first target, even though it was known that Genentech was also trying to clone the TPA gene. Through the good offices of Roger Samet, CBC formed an alliance with Baxter Travenol, but in 1982, Genentech announced the first cloning of TPA, and the alliance, as well as CBC, collapsed.

The Laboratory had more success with a collaboration with Exxon Research and Engineering, which wanted to move into molecular biology. Exxon provided substantial funds in return for CSHL providing training and research facilities for six Exxon scientists. These funds were used in part for an addition to Demerec Laboratory as well as expanding the plant program. Unfortunately the Exxon–CSHL collaboration ended prematurely when oil prices dropped and Exxon had to rethink its plans. There were other deals with other companies through the 1980s, including Amersham, Monsanto, Pioneer Hi-Bred International, and Oncogene Science.

By far the most valuable of these early commercial endeavors was the patent for cotransfection filed by Columbia University, naming Richard Axel, Saul Silverstein, and Michael Wigler as the inventors. CSHL began to receive royalties in 1991 after John Maroney's successful negotiations with Columbia.

*After an Absence of 50 Years, Neuroscience Returns
Briefly to Cold Spring Harbor*

Neuroscience research had been carried out during the Biological Laboratory's biophysics period, from the late 1920s to the early 1930s, albeit mainly by summer visitors. It had faded away as research focused on bacterial and viral genetics and, later, molecular genetics. However, once the genetic code had been cracked, it seemed that there was little more to do in molecular genetics other than dotting "i"s and crossing "t"s. Several molecular biologists, notably Francis Crick, Seymour Benzer, Gunther Stent, and Marshall Nirenberg, looked to neuroscience as the next great challenge and changed fields. Watson suspected that this movement of molecular biologists to neuroscience would continue and that there would be a need for courses to retrain the former in the latter. In the 1969 Annual Report he wrote of "applying to several agencies and foundations for funds" to support a neurobiology program.

Watson's applications were successful, and the Alfred P. Sloan Foundation provided a $450,000 grant to support a neurobiology course for five years, beginning in 1971. The funds were also used to renovate the 1914 Animal House, the building where McClintock had worked in the 1940s. It was renamed for her. Two courses were given in 1971 with a stellar cast of instructors and lecturers including Max Cowan, Zach Hall, David Hubel, Jim Hudspeth, Eric Kandel, Steve Kuffler, John Nicholls, Roger Sperry, and Torsten Wiesel. An important development that added much to the neuroscience courses was the renovation of Jones Laboratory. The by-then 80-year-old building was transformed, the interior walls removed, and four "pods" sheathed in aluminum installed. Each pod was self-contained, climate-controlled, and vibration-free, so that delicate equipment could be used. The neuroscience course program went from strength to strength, and by 1978 there were no fewer than 11 laboratory and lecture courses in neurobiology.

The renovated Jones Laboratory with the aluminum "pods," 1975.

It was not until November 1978 that neuroscience became a research division, albeit with only one member Birgit Zipser. Gunther Stent had promoted the leech as a research organism, and Zipser was carrying out electrophysiology of cells in the leech ganglion. By 1979, she was collaborating with Ron McKay, then in the Tumor Virus division, to raise antibodies against leech ganglion cells. The following year McKay moved to

Ron McKay, 1983.

Birgit Zipser and Susan Hockfield in Robertson House, 1980.

join Zipser, and Susan Hockfield was recruited. The three were successful at generating monoclonal antibodies that recognized individual cell types in the ganglia, but the collaboration was regrettably short-lived. Although Jones Laboratory had been renovated for the neurobiology group, it had to be vacated each summer to make room for the courses and bench space had to be found elsewhere. In 1984, Zipser and McKay left for environments more favorable for hard research, followed in 1985 by Hockfield, who went first to Yale and then became the first woman President of MIT.

Although the continuing highly successful neurobiology courses ensured that neuroscience was never completely absent from the Laboratory campus, it would be seven years before neuroscience was again a research area at Cold Spring Harbor.

Watson Goes to Washington

Although Dulbecco's speech at the dedication of the Sambrook Laboratory was perhaps the first public reference to a Human Genome Project (HGP), the idea of an HGP had first been floated by Robert

Sinsheimer at the University of California, Santa Cruz. The funding he was seeking did not come through, but the Department of Energy (DOE), looking for new uses for its National Laboratories, convened a meeting in March 1986 to review whether an HGP was feasible. That the DOE was interested in a project that seemed more relevant to the NIH was one of the factors that sparked the vigorous debate at the 1986 Symposium. Later that year, Charles DeLisi reallocated $5.3 million of DOE funds for an HGP beginning in 1987. This roused the NIH from its lethargy, and a flurry of meetings and reports, including ones from the National Research Council and the Congressional Office of Technology Assessment (OTA), recommended that the NIH should develop its own HGP. It was in the OTA discussions that Watson remarked in an exchange with John Sulston about who should run a HGP:

> Well, I couldn't think of a job I'd like less.... But I just have a feeling that someone has to keep track of the whole thing over the long period of time with a certain degree of intensity, it can sort of slip down and be run as a sort of B effort instead of an A effort.

Watson was never one to let an important project slip into B grade status and it did not surprise many people that when Jim Wyngaarden, NIH director, established an Office of Human Genome Research in September, 1988, Watson became its Associate Director. Following a Banbury Center meeting in 1989, a Memorandum of Understanding reconciling the activities of the NIH and DOE was signed and the publically funded HGP was under way. In October 1989, the NIH Office was upgraded to the Center for Human Genome Research so that Watson now had formal grant-giving powers. The job meant, he said, that he was spending more time in Washington than he had intended, but he was fortunate that the Laboratory now had senior management of "high competence" who could be relied on to "run their own shows and do so well."

James D. Watson speaking at the NIH, circa 1992.

Watson had from the beginning sought to allay fears that the authority and funds he wielded at NIH might entice him to Washington permanently. Shortly after his appointment, he wrote in the Annual Report that he had achieved a goal that he had had for many years—restoring Bungtown Road to its rural beauty by having all the power and telephone lines put underground. Now that it was done, "No one need further imagine that the corridors of real power will tempt me." He did return to CSHL in 1992 but under circumstances no one had predicted.

DAVID BEACH, CELL CYCLE CONTROL, AND CANCER

Summary

David Beach's protean contributions to research on cell-cycle control mechanisms in the 1980s and 1990s shed light on fundamental processes at the heart of one of biology's great mysteries and helped prepare the way for the development of novel anti-cancer therapies.

The Research

Before arriving at Cold Spring Harbor in 1982, David Beach was a postdoctoral investigator with Sydney Shall at the University of Sussex, U.K. Beach's skill in genetic manipulation of yeast led to a collaboration with Paul Nurse. The future Nobel laureate brought Beach into the thick of cell-cycle research, as they, along with many others, probed the molecular relationships between various cell-division-cycle *(cdc)* mutant genes in *Saccharomyces cerevisiae* and *Schizosaccharomyces pombe,* the model eukaryotes better known, respectively, as budding and fission yeasts. By the end of the 1980s, dozens of *cdc* genes had been cloned using complementation analysis, and *cdc* gene sequences had been deduced. Homologies and functional equivalencies were established over the immense range of species from marine invertebrates to yeast to humans.

Amid a frenzy of paper-writing, the teams Beach assembled at Cold Spring Harbor Laboratory would demonstrate by 1989 (further to intricate genetic studies in fission yeast which identified the relationship between cdcl3 [cyclin] and cdc2) that the two cyclins then known, designated A and B, associated (in distinctive ways) with protein kinase cdc2 and acted to regulate it. In complex during G_2, the pair formed MPF, the formerly elusive "mitosis promoting factor." No longer mysterious was cyclin fluctuation during the cell cycle, in contrast with the relative constancy of cdc2 and other species of what would later be called cyclin-dependent kinases, or Cdks. Within individual cycles, cyclins were synthesized and later marked for proteolytic degradation, one of the requirements for advancement through cell-cycle checkpoints.

Understanding not only that cyclins were necessary for a cell to enter mitosis, but also that their absence was a condition for mitosis to end hinted at one way in which, as many had long suspected, cell-cycle dysregulation might contribute to oncogenesis. The discovery of G_1 cyclins in

David Beach, 1994.

Continued

yeast set Beach and other investigators in search of their equivalents in mammalian cells. After comparing protein sequences, Beach and Charles Scherr, of St. Jude's Children's Hospital, realized that they had found the same gene, named by Beach as D-type cyclins. In 1992, Scherr discovered CDK4, the key catalytic partner of D-type cyclins, which phosphorylated the retinoblastoma (Rb) tumor-suppressor protein. The next year, Beach found p16INK4a, a tumor suppressor that inhibited the cyclin D-CDK4 complex. He had previously identified p21, another tumor suppressor.

Significance

From this work, the picture of a route to cancer—the Rb pathway—emerged, and a way around it. Extracellular mitogens triggered receptor-dependent cyclin D synthesis, allowing activation of CDK4, the phosphorylation of Rb, and entry into S phase. In response to stress, p16 was induced to inhibit CDK4, effectively suppressing its oncogenicity. This provided a starting point for industry to generate a CDK4 inhibitor. Early attempts failed, but in 2015, the FDA granted breakthrough-therapy approval to palbociclib (Ibrance), a first-in-class CDK4 and 6 inhibitor, for use in metastatic breast cancer (postmenopausal ER+/HER2−) in combination with letrozole.

The Scientist

David Hugh Beach was born in London on May 18, 1954. He received a B.A. from the University of Cambridge in 1975, and in 1977, he was awarded a Ph.D. by the University of Miami (Florida). A year after an influential summer at the Marine Biological Laboratory at Woods Hole, he began postdoctoral studies at the University of Sussex, U.K. In 1982, he arrived at Cold Spring Harbor as a postdoctoral investigator and quickly rose through the ranks, becoming a Senior Staff Scientist in 1988 and tenured scientist in 1992. He was named a Howard Hughes Medical Institute Investigator in 1990 and was elected a Fellow of the Royal Society in 1996. He has been awarded the Eli Lilly and Company Research Award (1994) and the Bristol Myers Squibb Award for Cancer Biology (2000). Since his departure from CSHL in 1996, Beach has held positions in London at the Blizard Institute, Barts, and The London School of Medicine and Dentistry, where he is Professor of Stem Cell Research.

— *P.T.*

WORKING OUT HOW DNA REPLICATES

Summary

DNA replication occurs through a carefully orchestrated, intricate dance between dozens of molecules. Bruce Stillman, now President and CEO of CSHL, has identified many of these proteins, illuminating how the genome is faithfully copied each time a eukaryotic cell divides.

The Discovery

In 1953, Watson and Crick first described the structure of DNA as a double helix. It was a watershed event, at least in part, because the structure implied how replication might occur. Since then scientists around the world have worked to build a detailed mechanistic picture of the events and molecular players that copy DNA.

Bruce Stillman's research has been instrumental in uncovering many of the proteins that control eukaryotic DNA replication. In the 1980s, Stillman used Simian Virus 40 (SV40) as a model for replication and developed an in vitro assay that coupled DNA replication with chromatin assembly, the process of assembling proteins on the newly copied DNA that make up chromosomes. He used this biochemical system along with genetic techniques to discover many of the proteins that are required for SV40 genome replication. In particular, he identified a number of DNA replication proteins and chromatin assembly factor CAF-1, a protein that is essential for nucleosome inheritance and the preservation of epigenetic information from one generation to the next. Ultimately, the work enabled Stillman and his team to fully reconstitute SV40 DNA replication with purified human proteins.

In parallel, Stillman's laboratory used budding yeast, *Saccharomyces cerevisiae*, to identify key proteins that determine the location of replication initiation sites along cellular chromosomes. In the late 1980s, scientists had isolated short stretches of DNA, known as autonomously replicating sequences (ARSs), that enabled *S. cerevisiae* to maintain nonchromosomal segments of DNA. Stillman and Stephen Bell, a postdoctoral fellow in the laboratory, realized that these sequences must recruit proteins that enable DNA replication, making them so-called "origins" of DNA replication. Stillman and Bell fractionated yeast cell extracts and incubated them with an ARS sequence to identify proteins that bound specifically to the origin. They isolated a group of proteins that they termed the origin recognition complex (ORC). This complex marks the site of replication initiation in eukaryotes across evolution, from yeast to humans.

The identification of the ORC was a breakthrough in the field of DNA replication. It enabled Stillman and others to discover all of the essential proteins required for replication initiation and DNA unwinding and eventually reconstitute initiation of chromosome replication with purified proteins. Although Stillman's laboratory continues to explore DNA replication, his team has also identified novel roles for replication proteins, such as ORC, in the processes that ensure accurate chromosome segregation during mitosis.

Significance

The instructions for life are stored within the billions of nucleotides that make up an organism's genome. Copying this information is one of the most fundamental steps in cell division, and errors can be catastrophic. Over the last 30 years, Stillman and others have identified the proteins and events that replicate DNA and control the process. This

Continued

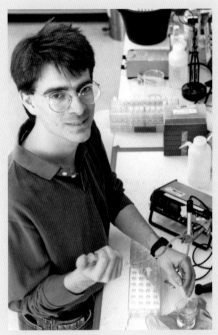

Stephen Bell in Bruce Stillman's laboratory, 1991.

knowledge offers great insight into normal cell proliferation control as well as diseases that are caused by errors in DNA replication, including cancer and rare genetic disorders.

The Scientist

Bruce Stillman was born in 1953 in Melbourne, Australia. As an undergraduate at the University of Sydney, he was fascinated by advances in recombinant DNA technology. His interest spurred him to attend graduate school at Australian National University where he studied DNA replication in human adenovirus.

Upon his graduation in 1979, Stillman came to Cold Spring Harbor Laboratory as a postdoctoral fellow. His work was bolstered by the many other researchers here who were studying how cell proliferation is controlled. Stillman was rapidly promoted to Professor. In 1994, Stillman became Director of CSHL and in 2003 he assumed the role of President, which he continues to hold today. Stillman lives in Airslie House on campus with his wife, Grace Stillman.

— *J.J.*

THE HARLOW LAB IDENTIFIES THE RETINOBLASTOMA GENE AS A TUMOR SUPPRESSOR

Summary

Tumor-suppressor proteins in normal cells regulate aspects of the cell cycle, controlling cell proliferation. If these genes are mutated, cancers develop. Ed Harlow's laboratory showed that cancers can also arise when tumor-suppressor proteins are inactivated by binding with proteins from tumor viruses, in this case the adenovirus E1A protein binding the protein product of the retinoblastoma (*Rb*) gene.

The Research/The Discovery

Physical interactions between viral oncoproteins and cell proteins were already known. For example, David Lane and Lionel Crawford at the Imperial Cancer Research Fund (ICRF) in London and Daniel Linzer and Arnie Levine at Princeton University had found that p53 bound to SV40 T antigen. Other such complexes included cellular p53 with the SV40 large T antigen, cellular src with the polyoma virus middle T antigen, and p53 with the adenovirus E1B protein. However the significance of these interactions was not clear.

Harlow became interested in these issues while a graduate student at the ICRF, making monoclonal antibodies to the SV40 T antigen and p53 protein. When he moved to CSHL, he focused on adenovirus, a DNA tumor virus, concentrating on the E1A protein, which is produced soon after the virus infects a cell. In the course of characterizing monoclonal antibodies against E1A protein, it became clear that the protein was binding to cellular proteins.

To determine what these proteins were, Harlow used these monoclonal antibodies to precipitate the E1A-cell protein complexes from extracts of adenovirus-infected cells, treated the extracts so that the proteins dissociated from each other, and examined the proteins by electrophoresis. Of the several unknown proteins that bound to E1A, one with a mass of 105 kDa was produced at high levels and became a focus for further characterization. As Harlow and his graduate students Peter Whyte and Karen Buchkovich were studying the mysterious 105-kDa protein, Wen-Hwa Lee (then at the University of California, San Diego) published a paper describing protein product of the Rb tumor-suppressor gene, a nuclear phosphoprotein with DNA-binding activity. It had a mass of 110–115 kDa, sufficiently similar to the Harlow protein to make it a candidate for the 105-kDa protein.

Harlow was in London when he read the Wen-Hwa Lee paper and was gratified to find that Whyte and Buchkovich had also made the connection. Immediately experiments began to determine if the Harlow 105-kDa protein and the Rb protein were the same, and indeed they were. Further work showed that the Rb protein normally acts as a brake controlling the progression of a cell from G_1 into the S phase of the cell cycle. When the viral E1A protein binds to pRB, the tumor-suppressor protein is inactivated, and cells enter S phase of the cell cycle to begin DNA synthesis and transformation.

Significance

The E1A- Rb protein complex revealed a new mechanism by which oncogenic viruses cause cancer. It was the prototype for subsequent discoveries: SV40 large T antigen and papillomavirus E7 proteins also inactivated Rb protein, while p53 was found to be a target of the oncogenic proteins of SV40, adenovirus, and human papillomaviruses. These protein interactions provided useful approaches to study the cellular proteins themselves and led to work that helped us to understand the functional roles for tumor-suppressor proteins in controlling tumor development.

Continued

The Scientists

Ed Harlow did his undergraduate work at the University of Oklahoma and earned his Ph.D. at the ICRF Laboratories (now Cancer Research UK). He came to the Sambrook Laboratory facility at Cold Spring Harbor Laboratory in 1982 to continue the research he had begun at ICRF and also to set up a monoclonal antibodies facility. Harlow moved in 1991 to Massachusetts General Hospital and Harvard Medical School, where he is Head of the Department of Biological Chemistry and Molecular Pharmacology and Virginia and D.K. Ludwig Professor of Cancer Research and Teaching. He has also served as Associate Director for Science Policy at the National Cancer Institute. Harlow is a member of the National Academy of Science and Institute of Medicine and a Fellow of American Academy of Arts and Sciences and has received the American Cancer Society Medal of Honor. He is an author with David Lane of *Antibodies: A Laboratory Manual* (1988) and *Using Antibodies* (1999).

After receiving his B.Sc. from the University of British Columbia in 1980 and M.Sc. in 1983, Peter Whyte earned his Ph.D. from the State University of New York 1987. He is an Associate Professor at McMaster University. Karen Buchkovich received her B.S. degree from Pennsylvania State University and from 1984 to 1990 was a graduate

Ed Harlow and Abhjeet Lele, Undergraduate Research Program, in 1986.

student in the Harlow lab, earning her Ph.D. at the State University of New York at Stony Brook. She has since held positions at the University of Illinois at Chicago Cancer Center and Moravian College.

— *J.C.*

16

A Change of Leadership

On July 6, 1990, the Laboratory hosted a daylong party in celebration of the 100th anniversary of the Biological Laboratory and the various incarnations that followed. Organized by Susan Cooper, CSHL librarian and archivist, more than 1500 people came to enjoy a picnic on Blackford lawn while the Old Bethpage Village Restoration Brass Band serenaded them with music that the first students of the Biological Laboratory might have heard. Four sailing ships were moored out in the Harbor to remind the partygoers of Cold Spring Harbor whaling history, while two tableaux replicated the scenes in photographs of an early class on the steps of Jones Laboratory and in the *Rotifer,* a small steam-powered launch. And the party was brought to a close with a display of fireworks by the Grucci family.

Indeed, there was much to celebrate, not least, given the crises over the years, that there was still an institute at Cold Spring Harbor. By 1990 the Laboratory was stronger than at any time in the previous 100 years. The Robertson Research Fund provided a substantial, stable income to support research; Laboratory scientists were successful in getting grants; the Meetings and Courses programs were attracting thousands of scientists each year; the Banbury Center and the DNALC were well established; and CSHL Press books were in demand and its journal *Genes & Development* was highly rated.

The move from tumor viruses to eukaryotic cells meant that the Laboratory's work continued to be at the cutting edge of cancer research. Plant science was also thriving, not least because of the Laboratory's

A clown entertains children at the Laboratory's 100th celebration party.

Reenactments of two scenes from the Biological Laboratory.

scientists being early players in the project to sequence a plant genome. In 1991, a new neuroscience program began that rapidly became a major component of the Laboratory's research. The standing of the Laboratory received a very public endorsement when Rich Roberts shared the 1993 Nobel Prize in Physiology or Medicine. And to ensure that CSHL research embarked on the next 100 years in confidence, a major fund-raising campaign, the Second Century, was initiated.

If there was one cloud on the horizon it was that Watson was commuting between Cold Spring Harbor and Bethesda, Maryland, trying to do two jobs at once, directing CSHL and the National Center for Human Genome Research. As early as January 1991, Watson had told the HGP Advisory Council that the project would soon need a full-time director, but the manner in which this came about just 15 months later was not a cause for celebration.

Neuroscience Returns as a Major Research Theme

The neurobiologists Zipser, McKay, and Hockfield departed for greener pastures, in part because they had had to vacate Jones laboratory each summer to make way for the summer neuroscience courses. Watson resolved that he would not recruit more neuroscientists until he could provide them with a new laboratory dedicated to neuroscience. He set up a small committee of trustees—Eric Kandel, John Klingenstein, William Robertson, and Charles Stevens—to explore how this could be done. The committee's advice was favorable and the Laboratory's longtime architects, Bill Grover and Jim Childress of Centerbrook, drew up ambitious plans for a two-part building on the hill to the west of Bungtown Road. One part would be laboratories and the second would provide accommodation for meetings and course participants, replacing the primitive, 30-year-old Page Motel. The project would be by far the largest ever undertaken at Cold Spring Harbor and would require an equally ambitious fund-raising effort.

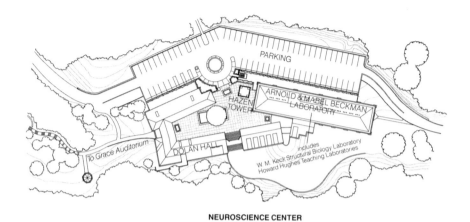

NEUROSCIENCE CENTER

Plan of Dolan Hall and the Beckman Neuroscience Center.

The Page Motel.

Early in 1987, the James S. McDonnell Foundation led the way with $1 million, and in July 1987, the Howard Hughes Medical Institute added $7 million. In 1988 the Beckman Foundation contributed $4 million; the W.M. Keck Foundation followed with $2 million. Mrs. Lita Annenberg Hazen donated $1 million and the Dolan Family Foundation contributed $2 million to the cost of the residential part of the project. The Landeau family donated $1 million, and, together with contributions from other Foundations and from members of LIBA, the necessary $21.5 million was raised by June 1990. Work on the building had begun in 1988 and it was ready to be occupied in 1991.

The question was: What areas of neuroscience should be the focus of research in the Arnold and Mabel Beckman Laboratory? The major challenge in neuroscience was cognition—learning, memory, consciousness—and of these, learning and memory were the most experimentally tractable. In 1974 Chip Quinn, then in Seymour Benzer's laboratory at Caltech, developed a reliable method to measure olfactory learning and memory in *Drosophila*. In the following years he and others isolated *Drosophila* mutants, such as *amnesiac, dunce,* and *rutabaga,* which had

The completed Dolan Hall and Beckman Neuroscience Center. The former is in the foreground in light colored brick and latter stretches to top right in dark brick. The Lita Hazen tower is obvious.

defects in learning and memory. Tim Tully, who had refined Quinn's technique, came to CSHL to continue searching for genes involved in learning and memory. Ron Davis had been working on the molecular characterization of the *dunce* mutation at Baylor College of Medicine. He continued that work as well as studying *rutabaga*. The last of the senior appointments was that of Hiroyuki Nawa, who worked on the development of neurotransmitters and how this changes during learning. The following year Alcino Silva and Yi Zhong joined the Beckman neuroscientists, who by then constituted a major research effort at CSHL.

It was most appropriate then that the dedication of the Beckman Neuroscience Center took place during the 1990 Symposium on *The Brain*. It was a remarkable meeting, not least because of the wonderful Dorcas Cummings Lecture given by Francis Crick, an event so popular that many extra seats had to be added to the stage area.

The Beckman building also provided a home for the X-ray crystallography group, which had been established in 1986 with Jim Pflugrath

Alcino Silva with Susumu Tonegawa at a meeting at the Banbury Center (1996).

(Left to right) Carla Shatz, Friedrich Bonhoeffer, Colin Blakemore, Martin Raff, and Paul Patterson at the 1990 Symposium on The Brain.

and John Anderson. The laboratories were in Demerec Laboratory, while offices and computers were in Hershey. This separation came to an end when, aided by a grant from the W.M. Keck Foundation, the group moved into purpose-built facilities in the lower level of the Beckman building.

Friends of the Laboratory—Neighbors and Trustees

Friends of the Laboratory played an important role in raising funds for the neuroscience center, both by introducing the Laboratory to potential donors and through their own personal contributions. Such support has been crucial ever since 1923, when the Long Island Biological Association (LIBA) was formed to take over the Biological Laboratory from the Brooklyn Institute of Arts and Science (see Chapter 5). Funds raised by the LIBA provided unrestricted funding for the Biological Laboratory and helped fund the construction of new and the renovation of old buildings.

LIBA members were drawn principally from the local community, and what a community it was (and still is). The stretch of the north coast of Long Island running from Sands Point in the west to Lloyd Harbor in the east was early one of the wealthiest areas in the northeast. With an easy commute to New York City by rail, beautiful land, and a lovely coastline, it became a favorite place for wealthy financiers, industrialists, and others to build their homes. The Marshall Fields' estate at Caumsett was the most spectacular, but there were other estates hardly less impressive.

Following the creation of CSHL in 1962, the LIBA continued to support the Laboratory as before and even more successfully. In 1972, for example, Ed Pulling, Chairman of the LIBA, led a campaign to raise $250,000 to build a new addition to James Laboratory (the Sambrook extension [page 260]) and to winterize Blackford Hall. A few years later, Pulling led another LIBA campaign for $225,000 to build a new Williams House, the original decrepit beyond repair, and a further $200,000 to purchase land still owned by the Carnegie Institution of Washington. And, a gift appreciated by hundreds of thousands of scientists, the LIBA contributed $600,000 toward the cost of the new Oliver and Lorraine Grace Auditorium. By the mid-1990s, Cold Spring Harbor Laboratory

Ed Pulling (far left) and Walter Page (second from right), both long-term supporters of the Laboratory and former chairs of LIBA, with Jane Page and Joe Sambrook (far right) at the dedication of the Sambrook Laboratory.

Association (CSHLA), as LIBA had become in 1991, was contributing more than $500,000 annually to Cold Spring Harbor science.

The LIBA had one official member on the CSHL Board of Trustees (as did the Wawepex Society), but the LIBA had a larger presence on the Board as several of the individual trustees were also LIBA members. The scientist members of the Board represented the 10 contributing institutions that had provided financial support to the Laboratory, and included scientists from other institutions as needed. Like the LIBA, the trustees played a major role in fund-raising campaigns. With the Laboratory's centennial in 1990 approaching, a committee was set up to raise money for the Second Century Campaign. Led by David Luke, the chairman of the Board of Trustees, and with the support of CSHLA members, foundations, and companies, it achieved its goal of $44 million by 1991 and went on to reach $51 million. Together with the Infrastructure Fund, led by trustee Bill Miller, these funds were used in part to renovate two of the oldest buildings on the campus—McClintock Laboratory and Blackford Hall. Centerbrook Architects designed a third story for McClintock Laboratory while retaining the charm of the building, as well as carrying out extensive changes internally. The changes at Blackford were even more drastic. Built in 1908, it had long been inadequate to deal with the increased numbers of staff and participants in meetings, and for many years Jim Hope, director of Culinary Services, and his staff had coped admirably with cramped conditions and inadequate facilities. The new addition to Blackford doubled the dining capacity to 400 and provided new space for kitchens and food preparation, not to mention a larger and more attractive bar.

Relationships between the Board and the director have not always been smooth. As recounted in Chapter 13, John Cairns found Ed Tatum obstructive and unsympathetic to Cairn's attempts to establish CSHL as a new institute, risen from the ashes of the Biological Laboratory and the Department of Genetics. Indeed, Cairns suspected that Tatum was

David Luke with Bill Miller, leaders of fund-raising campaigns and former chairs of the Board of Trustees, at the Banbury Center in 1990.

The balcony of Blackford Hall is demolished to make way for the new extension and balcony in 1990.

hoping that CSHL would fail so that it could be taken over by Rockefeller University. Matters came to a head when, during the turmoil attendant on Cairn's impending resignation, Walter Page, Vice-Chairman of the Trustees, invited Tatum to dinner and told him that Tatum had to relinquish his chairmanship of the Board. (The Pages had a long association with Cold Spring Harbor. Arthur Page had been a founding director of the LIBA in 1924 and served on the LIBA board until 1958 when Walter became President. He retired from the board in 1986.)

The most celebrated of Watson's disagreements with his board came over the marina on the shore of the inner harbor, opposite the Laboratory. In 1960, Demerec and Chovnick had thwarted an attempt to develop the inner harbor, but in the early 1970s, Arthur Knutson, owner of the Whaler's Cove marina directly across from the Laboratory, proposed expanding it from 50 boat slips to several hundred. He was, however, willing to sell it, and the matter was discussed at the June 1972 trustees' meeting. Watson recommended that the Laboratory buy the marina to ensure that the beauty of the Harbor was preserved for scientists and local community alike. However, Ed Pulling and Walter Page objected to using funds raised

A view of the inner harbor looking north, the sand spit at the top *and the* Laboratory *lower left.* Whaler's Cove *is the small marina on the* right. *In this photo from the early 1960s there are some 40 boats at anchor. If the Laboratory had not bought the marina, there would now be several hundred boats moored in the inner harbor.*

to support research for the purchase. In the bland words typical of official minutes, it is recorded that

> Dr. Watson again stated his very strong conviction that the purchase was an absolute necessity for the future of the Laboratory. He stated that he viewed the matter as a vote of confidence and that because of the strength and emotional nature of his convictions on the subject, he would prefer not to participate in the remainder of the discussion and departed.

Barney Clarkson from the Sloan-Kettering Institute and secretary (later chair) of the board tells it as it was: "Jim exploded … and stormed out of the room." On further reflection, including the realization that income from the marina could be used to pay back a loan for the purchase, the trustees authorized the purchase. It was completed the following year and our scientists and visitors still enjoy the beauty and tranquility of the Harbor.

Members of the CSHLA and the trustees have played an important role beyond raising funds by introducing the Laboratory to friends and colleagues, who in their turn become CSHLA members and trustees. Perhaps

Barney Clarkson (right) with Ernst Mayr at the 1987 Symposium.

the most consequential of such introductions was made by Ed Pulling when he phoned Watson to tell him that a local neighbor wanted to make a gift to the Laboratory. The neighbor was, of course, Charles Robertson.

Research

Carol Greider, circa 1990.

Carol Greider had come as a Cold Spring Harbor Fellow in 1988 and become a staff member in 1989. She had continued to work on telomeres, the sequences at the ends of chromosomes that protect the ends from damage. Because of the way DNA replicates, telomeres shorten each time a cell divides. There is an enzyme, telomerase, which Greider and Elizabeth Blackburn discovered, that repairs telomeres, but this enzyme is only active in cells like stem cells. Normal cells growing in cell culture become senescent after about 50 cell divisions, and Greider, together with Bruce Futcher and Calvin Harley, showed that the telomeres of such cells shorten as the cells age in tissue culture, suggesting that telomere shortening is a contributing factor in senescence in the whole organism. In contrast, cancer cells growing in culture do not senesce, and Greider et al. showed that these have an active telomerase. Indeed, it was soon shown that cells in tumors, and not simply cancer cells growing in cell culture, also express telomerase, showing that reactivation of telomerase is an important step in the development of cancer.

During this period several laboratories were working on the RAS and related pathways in cancer, determining the proteins involved in taking a signal from the cell surface to turn on genes in the nucleus. Mike Wigler, Dafna Bar-Sagi, Linda Van Aelst, and Mike Gilman all made contributions to this work. Nick Tonks discovered a new class of proteins, the tyrosine phosphatases, which are involved in signaling pathways, and disturbances in tyrosine phosphatase function have been shown to be involved in cancer as well as other disorders such as diabetes, auto-immune disorders, and neurological diseases. These enzymes are the subject of intense study as therapeutic targets. David Beach continued to work on the cell cycle and discovered three proteins—cyclin D, the partner of the kinase CDK4, and its inhibitors p16 and p21—that play key roles in controlling cell division. Alterations in p16 have been shown to be important in a variety of cancers. In 2015, a major new cancer drug targeting cyclin D–CDK4 was approved for use in breast cancer.

In theory the simplest way to find the difference between, say, a normal cell and a cancer cell is to compare the genomes of the two and look for genes that differ. However, the size and complexity of the human genome made this approach untenable until Mike Wigler and Nikolai Lisitsyn developed a technique called representational difference analysis (RDA). RDA proved to be very useful for finding differences (mutations) between normal and cancer tissues and between harmless and pathogenic variants of bacteria.

The neuroscientists in this period were working hard at dissecting the genetic and molecular mechanisms involved in learning and memory. Ron Davis had left in 1993 to take up an endowed chair at Baylor College of Medicine, but the group in Beckman was augmented by Jerry Yin, who worked on the fruit fly, and by Alcino Silva, who upped the stakes by working on mice. The work of the Beckman group helped

Linda Van Aelst, circa 1997.

Nick Tonks, 1993.

*Richard Roberts receiving his
Nobel medal and citation from King
Carl XVI Gustaf, 1993.*

establish the importance of a protein called CREB in the development
of long-term memory.

*Rich Roberts and Phil Sharp Share the 1993 Nobel Prize
in Physiology or Medicine*

Rich Roberts had left CSHL in 1992, and in the following year the welcome news came from Stockholm that he was to share the Nobel Prize with
Phil Sharp for the discovery of splicing. It was generally felt that the award
was long overdue, as splicing was arguably the most significant discovery

in molecular biology since the deciphering of the genetic code. As so often happens with Nobel Prizes, there was disquiet that worthy candidates, in this case Susan Berget and Louise Chow who had played key roles in the work at MIT and CSHL, respectively, had been excluded. Unfortunately there will always be a "41st chair" for any honors with a restricted number of recipients.

A Change in Leadership

By 1990 the HGP was well underway, but Watson was feeling an increasing strain in trying to both run a federal program that now had a $60 million budget and look after Cold Spring Harbor. What he needed, he wrote, was a "capable second." Bruce Stillman had been approached about becoming Director of the Virus Laboratory at the University of California, Berkeley, but unbeknownst to him, Watson had been exploring offering Stillman the position of Assistant Director. In September 1990, Stillman took on the post of Watson's deputy while Rich Roberts and Terri Grodzicker remained Assistant Directors for Research and Academic Affairs, respectively. However, Stillman's appointment made it clear that he was in line of succession to Watson, and in 1992 Roberts left to take up a full-time position as Chief Scientist at New England BioLabs.

Watson's return to Cold Spring Harbor Laboratory was not long in coming. Although the HGP was established, Watson was increasingly frustrated at the NIH and his relationship with Bernadine Healy, NIH director, was increasingly chilly. Matters came to a head over patenting of DNA sequences.

Craig Venter, then laboratory head in the National Institute of Neurological Disorders and Stroke, submitted a proposal to sequence not the whole genome but just those parts of it that code for proteins, cDNA sequencing. This would provide the sequence for the key elements of the genome, its genes, at a substantially reduced cost, but as this strategy

News & Comment

Genome Patent Fight Erupts

An NIH plan to patent thousands of random DNA sequences will discourage industrial investment and undercut the Genome Project itself, the plan's critics charge

AT A CONGRESSIONAL BRIEFING ON THE Human Genome Project last summer, molecular biologist Craig Venter of the National Institute of Neurological Disorders and Stroke dropped a bombshell whose repercussions are still reverberating throughout the genome community. While describing his new project to sequence partially every gene active in the human brain, Venter casually mentioned that his employer, the National Institutes of Health, was planning to file patent applications on 1000 of these sequences a month.

"I almost fell off my chair," says one briefing participant who asked not to be named. James Watson, who directs the genome project at NIH, did more than that, exploding and denouncing the plan as "sheer lunacy." With the advent of automated sequencing machines, "virtually any monkey" can do what Venter's group is doing, said Watson, who in one sentence managed to

conceivably lay claim to most of the human genes. This, they say, would undercut patent protection for those who labor long and hard at the real task of elucidating the function of the proteins encoded by the genes, thereby driving industry away from developing inventions based on that work. "No one benefits from this, not science, not the biotech industry, not American competitiveness," asserts Botstein, who also attended the hearing last summer and has since been trying to mobilize the genome community to oppose the idea.

The scheme's critics envision a mad scramble for patents. "If Craig can do it, so can the UK," Watson told *Science*. Indeed, Bodmer has warned that if Venter contin-

"No one benefits from this, not science, not the biotech industry, not American competitiveness."
—Botstein

pany are wrong about the effect on industry, insisting that patent and license protection will help—not hinder—technology transfer. What's more, they say that, given the uncertainty over whether the clones are even patentable, they were just doing what is prudent by filing now.

Science article focuses on the patenting controversy.

would miss the sequences that control gene expression, his proposal was turned down. Nevertheless, Venter was able to find enough money to sequence cDNAs isolated from the human brain and suggested that NIH should patent the sequences. Uproar ensued and Watson vehemently and publicly opposed patenting gene sequences as "sheer lunacy." However, Healy supported patenting and must have decided that Watson had to go.

The opportunity came when Watson's application for a waiver concerning his earnings outside NIH (salary at CSHL, income from books and investments) came up for review. This waiver had been granted routinely but on this occasion, Healy asked Jack Kress, the special counsel for ethics in the Department of Health and Social Services, to review the matter. Although Kress ruled that there was no conflict of interest, Watson regarded it as a vote of no confidence by Healy, and he resigned on April 10, 1992.

DEPARTMENT OF HEALTH & HUMAN SERVICES   Public Health Service

National Institutes of Health
National Center for Human
Genome Research
Bethesda, Maryland 20892

Building 38A, Room 605
(301) 496-0844
(301) 402-0837 (Fax)

April 10, 1992

At 1:45 p.m. today, Dr. James D. Watson, Director of the National Center
for Human Genome Research, met with National Institutes of Health Director
Dr. Bernadine Healy to present his formal letter of resignation, effective
immediately. The text of the letter follows:

Dear Dr. Healy;

I hereby resign as Director of the National Center for Human
Genome Research. I have considered it a great pleasure and
opportunity to have served at the National Institutes of Health in
this capacity. I remain firmly committed to the success of the
Human Genome Project. I hope and expect to continue to support
the project enthusiastically and, if called upon by my successor,
to advise NIH informally.

Sincerely,

James D. Watson

In his conversation with Dr. Healy, Dr. Watson recommended that a
scientist of the highest reputation and integrity be appointed as soon as
possible to succeed him.

*The press notice released by the
office of Bernadine Healy, Director
of the National Institutes of Health,
announcing Watson's resignation from
the National Center for Human Genome
Research, 1992.*

On Watson's return to CSHL, there was a not-unexpected reorganization. At its 1993 meeting, the Board of Trustees appointed Stillman
as Director, while Watson took the newly created position of President.
As director, Stillman took responsibility for research and meetings
and courses, while Watson continued to look after the programs at the
Press, Banbury Center, and the DNA Learning Center and fund-raising.
Stillman took up his new position on January 1, 1994, and has followed

*Bruce Stillman with Jim Watson,
following Stillman becoming CSHL
Director and Watson President in 1994.*

the examples of his predecessors, Davenport, Demerec, and Watson, each of whom left his mark on both the science and the landscape of Cold Spring Harbor.

CAROL GREIDER AND TELOMERASE

Summary

In 1984, Carol Greider, while working in Elizabeth Blackburn's laboratory, discovered telomerase, an enzyme that adds a protective repetitive sequence to the ends of chromosomes. Her work has implications for numerous diseases and aging, and in 2009 Greider and Blackburn shared the Nobel Prize in Physiology or Medicine with Jack Szostak.

The Discovery

The DNA replication machinery is unable to copy the ends of chromosomes, so that DNA sequences at the ends are lost in each round of replication. How then is genetic information protected? In the 1980s, Elizabeth Blackburn's laboratory at the University of California at Berkeley identified long repetitive sequences at the end of chromosomes. Soon after, Carol Greider joined Blackburn's laboratory as a graduate student and set out to identify how these sequences, known as telomeres, are elongated.

Greider first discovered that the telomere is extended in a discrete unit of six nucleotides, TTGGGG. She devised a biochemical assay for the enzyme that adds this sequence to the ends of chromosomes and in 1984, she and Blackburn reported the discovery of telomerase. By the end of her graduate work Greider found that telomerase is composed of protein and RNA template.

In 1988, Greider arrived at Cold Spring Harbor as a CSHL Fellow. Her first goal was to clone the gene that encoded the RNA component of telomerase. In 1989, she and Blackburn published the work in *Nature*, together with a proposal for how the telomere repeat is made, later shown to be correct.

Significance

Because of their role in protecting the ends of chromosomes, telomeres are critical for genomic stability. Shortening of telomeres occurs during cell aging while cancer cells,

Carol Greider and Adrian Krainer, 1993.

which can divide endlessly, often have increased telomerase activity. Abnormalities of telomeres have also been found in human genetic disorders.

The Scientist

Greider chose to attend UC Santa Barbara for her undergraduate studies, where she quickly became enamored with molecular biology and carried out research through her senior year, ultimately applying to graduate school to work with Elizabeth Blackburn. On completing her Ph.D., she applied for a postdoctoral position at CSHL, but Bruce Stillman offered her an independent position as a CSHL Fellow.

In 1997, Greider moved to Johns Hopkins University in Baltimore, where she is the Daniel Nathans Professor and Director of Molecular Biology and Genetics program.

— *J.J.*

17

Into the 21st Century

At the time of the centennial in 1990, the Laboratory was undergoing a transformation that has continued to the present time. Bruce Stillman had become Assistant Director, focusing on research and the Meetings and Courses program, and in 2000, the CSHL Board appointed him Director and CEO to take charge of all activities. In 2003, the Board resolved the ambiguity of the Laboratory having a director and a president by Stillman becoming President while Watson took on the title of Chancellor. In 2001, Stillman reorganized the Laboratory's research groups, creating six divisions with a seventh, Quantitative Biology, added in 2008. Changes in the management structure of the Laboratory were needed as the number of scientists and operating budget continued to increase, and David Spector took on the role of Director of Research in 2007, following the departure of Winship Herr who was Assistant Director for Research from 1994 to 2005.

David Spector, circa 2006.

Even as the research grew, the Laboratory's commitment to education was unwavering. Grace Auditorium was proving to be a great success with meetings participants, and the Meetings and Courses programs extended beyond Cold Spring Harbor. The DNA Learning Center, too, expanded its activities to other states and countries, while the Laboratory became a Ph.D.-granting body with the opening of the Watson School of Biological Sciences in 1998.

In the past 25 years the Laboratory staff increased from 440 to 1200, and the inflation-adjusted total expenditures rose from $90 to $150 million. These increases were accompanied by significant changes in the landscape as new buildings were added to accommodate the new science.

The CSHL Facilities department and Centerbrook Architects carried out more than 25 projects—new buildings and major renovations—in this period. Even so, the pressing need for extra administrative space to support the growing number of people and programs at the Laboratory has forced the use of a site off-campus, at Syosset.

These changes could not be accomplished without, as Stillman put it, "…a very capable administration that nurtures science while minimizing obstacles." CSHL has been fortunate in having a series of high-level administrators who have not only been first-rate but who have felt an affinity for the style of the Laboratory and carried out their work with a light hand. The first was James Brainerd, soon to be succeeded in 1969 by Bill Udry, who served for 15 years. Morgan Browne took over from Udry in 1985, and he too was Administrative Director for 15 years, until 2000 when Dill Ayres became Chief Operating Officer. The Laboratory has benefited enormously from the long service of Udry, Browne, and Ayres.

Building Cold Spring Harbor

The transformation of CSHL in these years is most evident in the dramatic changes in the landscape of the west shore of the harbor. That these changes are in keeping with what has gone before and yet provide for the future is due to the hard work of many, and not least the good taste of Jim and Liz Watson, Bruce Stillman, and others who know that beautiful surroundings benefit residents and visitors alike. In 1995, the addition of the Laboratory to the New York State and National Registers of Historic Places was an acknowledgment of the dedication of Liz Watson and everyone else to conserving the Laboratory, while at the same time allowing it to flourish as a modern research institute.

Watson first came across the work of the architect Charles Moore when he and Liz were staying at Sea Ranch Condominiums in 1972. Moore established the architectural firm Centerbrook, taking its name

In 1995 Bernadette Castro (right), New York State Commissioner of Parks, Recreation and Historic Preservation, presented Liz Watson with the certificate proclaiming that the Laboratory is on the list of New York State and National Registers of Historic Places.

Condominium 1, Sea Ranch, Sonoma County (left) and the James Laboratory extension (right), both designed by Charles Moore.

from its host town in Connecticut. Centerbrook's first commission at CSHL was the striking renovation of Airslie House in 1974. Subsequently Centerbrook, represented most notably by Bill Grover and Jim Childress, has been involved in the design of virtually every new building and renovation at Cold Spring Harbor. The distinctive style of their designs, modern yet consonant with the styles of the buildings going back to the late 19th century, gives a consistent but varied style to the Laboratory's appearance.

Much of the construction and landscaping has been carried out by the Laboratory's own crews. Remarkably, there have been only two heads of Buildings & Grounds in 40 years. Jack Richards, a local contractor, came in 1970, taken on to complete the addition to James Laboratory. At that time, the department staff totaled 14; by 1997, the year that Richards retired, it numbered 80. Over that 27-year period, Richards and his staff were involved in 32 major projects, one of the last of which was a new home for the Buildings & Grounds department, aptly named the Richards building. Art Brings, Richards's deputy, took over, and has overseen a program as vigorous as that of his predecessor and a department with more than 130 employees. Brings has overseen some 25 major projects, all of which contribute to the events described in this chapter.

With the Beckman Neuroscience building and the Hazen Tower rising in the background, the Centerbrook and Laboratory teams pose for a photograph: (left to right) Bill Grover (Centerbrook), Jack Richards (Head, Buildings and Grounds), Morgan Browne (CSHL Administrative Director), Jim Childress (Centerbrook), and Jim Watson.

There is, sometimes, grumbling that the Laboratory spends too much on its buildings and grounds, but it is a delicate balance. On the one hand, the grounds and buildings should not look so dilapidated that a potential donor fears to put good money after bad nor so groomed that the donor thinks CSHL is so wealthy that no extra funds are needed. Looking at the Laboratory, whether from across the harbor or up close, it is clear that the balance has been achieved. But above all, beautiful buildings and beautiful grounds provide a wonderful environment for CSHL scientists and visitors.

Ballybung, Books, and Babies

Not all the new buildings were laboratories, and in fact, the first new building in this period was a house. The Director's house, Airslie, was built in 1806 and had been purchased by the Biological Laboratory in 1943 at a cost of $10,000. Occupied successively by Demerec, Cairns, and Watson, as it was to remain the Director's house and be occupied by the Stillman family, a new home was needed for the CSHL President. The Laboratory owned a property at the north end of the campus with a view across Long Island Sound to Connecticut. It had been part of the Nethermuir estate, the de Forest's grand house, which had been razed long before by Julia, Henry de Forest's widow. Here Jim and Liz built a house, Ballybung, with ample room for fund-raising and other events.

The 1904 building that had been in turn the Station for Experimental Evolution and the Department of Genetics had been used as the Library since the early 1950s. Susan Gensel (later Susan Cooper) came in 1972 as librarian and archivist (as well as head of Public Affairs), just at the time when there was a huge increase in research as the tools of molecular biology were taken up by researchers in many fields. Before Cooper left in 1997, the digital revolution had begun. At the time it was a major advance when the National Library of Medicine's MEDLINE database became available quarterly on a compact disk. The digital transformation of the Library continued when Mila Pollock, who came in 1999, greatly expanded online access to journals. The Library reinvented itself, and in a few years the transition from housing print copies of journals to providing CSHL scientists with access to electronic information sources was complete. Pollock and her staff also created the Institutional Repository of all publications by CSHL scientists, an invaluable resource.

The Library building had not received any significant attention for more than 60 years and was sorely in need of rejuvenation. In 2010 it underwent an extensive renovation and a major extension; initiated by a generous gift from Waclaw Szybalski, a reading room was added in the extension being named in his honor. The addition was also made possible by a gift from Genentech Inc., which helped create the Genentech Center for the History of Molecular Biology and Biotechnology. The new addition included offices, a handsome reading room, and, most importantly, a climate-controlled vault for the storage of the Laboratory's archives.

The Waclaw Szybalski Reading Room in the Library extension.

These archival materials had been stored in the Library attic, a space that had always been woefully inadequate for research and storage. Matters reached a crisis as Pollock made the Laboratory a major repository for personal archives relating to molecular biology. These include the papers of, among others, Hermann Muller, Jim Watson, Norton Zinder, Sydney Brenner, and Charles Yanofsky, all now housed in a state-of-the-art facility. The collections of Watson and Brenner have been digitized and are available online. In addition, the Genentech Center has been conducting oral histories with many of the contemporary leaders in molecular biology, neuroscience, and biotechnology. The Library promotes the use of its resources through the Sydney Brenner Research Scholarship and the Ellen Brenner Memorial Fellowship.

Just as the Laboratory has undergone a transformation, so the Library is no longer simply a place for journals and books; as well as a center for thinking, reading, and working, it is a place for hosting cultural events and meetings about the history of science.

A third building project was a day-care center. CSHL has thrived scientifically by recruiting bright, young scientists who stay for several

De Forest Stables as the new day-care center.

years and then move on to more senior positions elsewhere. In 1997, for example, the average age of the scientific staff was 35. Young people have young families with young children but scientists rarely have regular working hours, making it difficult to arrange child care. Carol Greider was chair of a faculty committee charged with recommending components of a potential sports facility, but she came back with the recommendation that a child-care facility should be built instead. Members of the Cold Spring Harbor Association, lead by Mary Lindsay and Wendy Vander Poel Russell, set out to raise sufficient funds. Set at the north end of the campus, the De Forest stables provided the ideal location. They had been used for accommodation, but now the U-shaped building underwent a beautiful conversion to provide rooms for infants, toddlers, and preschool children, and the area between the arms of the U was developed as a playground. It was most appropriate that the facilities were named for Mary Lindsay, a long-term Trustee of Cold Spring Harbor Laboratory.

The Meetings and Courses Program Flourishes

During his time as director, Jim Watson directed the meetings program with the aid of meetings coordinator Barbara Ward, and when Watson went to Washington, Terri Grodzicker helped oversee the programs as Assistant Director for Academic Affairs. When Bruce Stillman became Director in 1994, he initially focused on research and the meetings and courses programs. The latter had steadily increased so that by 1993 there were 25 courses and 15 meetings and it was clear that the program needed a full-time director in the same way that the CSHL Press, DNA Learning Center, and Banbury Center had their directors. One of Stillman's first actions was to recruit David Stewart from a biotechnology company in Cambridge, United Kingdom, to take the position of Director of Meetings and Courses. Grodzicker continued to play a key role as Assistant Director of Academic Affairs and combined these duties with her work as a research scientist and later as editor of *Genes & Development* at the CSHL Press.

Under Stewart's directorship, the number of meetings and courses continued to grow, so that by 2013 there were 30 meetings, with meetings now being held in early spring and late fall, far outside what had been the "traditional" meetings season. Even so, there were too many topics to fit in the Cold Spring Harbor calendar, and Stewart initiated a collaboration with the Wellcome Trust conference center at Hinxton, Cambridge, so that a small set of meetings alternated between Cambridge and Cold Spring Harbor. A far more significant venture began in 2008, when a wholly owned subsidiary of the Laboratory called Cold Spring Harbor Conferences Asia was established. Some 60 miles to the west of Shanghai, the Suzhou Industrial Park funded the construction of a hotel

David Stewart at the 1994 Symposium.

and conference center complex by the side of Dushu Lake. The inaugural meeting was the James Watson Cancer Symposium held in April 2010. Under the direction of Maoyen Chi, by 2015, 16 meetings were held in Suzhou. Now each year a total of some 12,000 scientists attend meetings and courses at Cold Spring Harbor and Suzhou.

The reach of the Laboratory's meetings is further extended by the internet, which came into its own in the 1990s with the development of Mosaic, one of the first web browsers and the one credited with driving the explosive growth of the World Wide Web. The Meetings & Courses programs make full use of the internet to disseminate the scientific research and techniques presented during the year. Since 2003, the Leading Strand site has hosted a variety of talks, interviews, and even electronic posters presented at CSHL meetings and courses; these are viewed by many thousands of scientists. In addition, there are online resources including open-access lecture videos such as keynote addresses at meetings and the Dorcas Cummings Lecture.

Although webcasts and other online resources have made possible the worldwide dissemination of CSHL meetings, they are unlikely to supplant the benefits of meeting colleagues face-to-face, whether to discuss results or seek job opportunities. But although the social events associated with a meeting are highly productive, for many years the facilities for poster sessions were inadequate. These were cramped and posters had to be distributed between several buildings. Fortunately, the 125th anniversary of the Laboratory was marked by the opening of the Nicholls Biondi Hall, a building dedicated to providing additional space for posters. Generously funded by Jamie Nicholls, chair of the Board of Trustees and her husband, Fran Biondi, it ensures that CSHL continues to be the best venue for the exchange of information and ideas.

Like the meetings, the number of courses increased in this period, reaching 32 in 2013, and they continue to introduce the newest techniques to students. An interesting example was the DNA microarray

The Nicholls Biondi Hall (2015) on the right *with Davenport House (1884) on the* left, *the construction of the two buildings spanning more than 130 years.*

course introduced in 1999 and led by Pat Brown, who had developed the technique only a couple of years earlier. The course was notable because the students built their own microarrayer as these were not yet commercially available. Over time, as the courses increased in number and the techniques increased in sophistication, new teaching laboratories

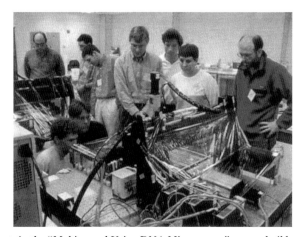

Students in the "Making and Using DNA Microarrays" course build an array.

were needed. The Howard Hughes Medical Institute (HHMI), which has long supported CSHL courses, provided the funding for a new building on the site of the old Hershey building. The new Hershey building has two large rooms for teaching computational biology courses, a wet lab that expanded space for new courses in experimental biology, and two laboratories that house shared research resources—microscopy and flow cytometry, which are two of the dozen shared research resources that provide very advanced technologies for research and courses at Cold Spring Harbor.

The Laboratory Opens Its Own Graduate School

The Laboratory has had graduate students for many years and the first to be awarded a Ph.D. was Bill Sugden, who in 1973 received his doctorate from Columbia University for his research on RNA polymerase from human cells. Subsequently, graduate students came principally from the State University of New York at Stony Brook, carrying out their research at Cold Spring Harbor but receiving a Stony Brook University Ph.D. Beginning in the early 1990s, serious discussions began about the Laboratory becoming a degree-awarding institution in its own right. It was not a decision to be taken lightly. How would the costs of students be covered? Would the faculty, who had come to CSHL without any intention of teaching, be willing to do so? In the end, funds were raised so that each student position was endowed and the faculty found that teaching was a good way of getting graduate students in their laboratories.

Winship Herr, who would become the first Dean, recounts that Watson discussed the idea of a graduate school at the November 1995 meeting of the Board of Trustees. Herr was rather taken aback when Watson, without prior consultation, proposed that Herr, whom Stillman had appointed Assistant Director for Research in 1994, should chair a subcommittee of the Board of Trustees to investigate the feasibility of such a graduate program. Eighteen months later, the Board resolved that the Laboratory should apply to the New York State Department of Education for accreditation as a Ph.D. degree–awarding institution, accreditation that was obtained in 1998. Board Chairman David Luke and Stillman then went to see Watson to convince him that the School should bear his name, and after much persuading, CSHL had a Watson School of Biological Sciences (WSBS).

Herr was joined in early 1999 by Lillian Clark Gann as assistant dean, and together they established the school as a reality, the first students joining in the Fall of that year. When Herr left for the University of Lausanne in 2004 he was succeeded as Dean by Lillian Clark Gann until she left for Cancer Research U.K. in London in 2007. Subsequently, CSHL faculty member Leemor Joshua-Tor was dean from 2007 to 2012, and the current Dean is Alex Gann.

From the beginning, the WSBS set out to distinguish itself from other graduate programs. The most notable feature is that students would aim to complete their Ph.D. in 4 years, this at a time when the median length of time for a Ph.D. in life sciences is between 6 and 7 years. Another special feature is a "two-tier" mentoring program: In addition to their research mentor (the doctoral thesis advisor in whose laboratory a student carries out their research), each student is matched with a faculty member whose duty is to look after the best interests of the student. Further, all course work is carried out in the first semester before the students start laboratory work, allowing them to concentrate exclusively on each activity rather than juggling them. And perhaps the single most important feature in speeding up the Ph.D. is that each student meets with his/her faculty thesis committee every 6 months, not annually or even less often as is traditional.

The School matriculated its first students in September 1999, and the first student to receive a Ph.D. was Amy Caudy in 2003, with five other

(**Left to right**) *Patrick Paddison, Ira Hall, Elizabeth Thomas, Niraj Tolia, Amy Caudy, and Emiliano Rial Verde at the first WSBS convocation, April 25, 2004.*

students joining her in the school's first graduation ceremony in spring of 2004. By all measures, the WSBS has been a success. It attracts outstanding applicants from throughout the world and competes for them, successfully, with top universities and institutes. Its students have published more than 300 papers from their research work, an average of more than three papers per student. Of the 80 students who graduated in the school's first 15 years, all did so within 6 years of starting their studies, with the average time to graduation being ~4.8 years. More than 80% of Ph.D. awardees have gone on to postdoctoral training in academic research laboratories, and of the students who graduated 5 or more years ago, about half are now in tenure-track faculty or independent research positions at major U.S. or international institutions. Other graduates are employed in nonresearch, science-related careers such as the pharmaceutical industry, consultancy, publishing, or biotechnology management.

Dick McCombie, circa 2007.

Doreen Ware, circa 2010.

Research at CSHL

It is impossible to cover the full depth and breadth of research at the Laboratory during the period covered by this chapter. Not only did the number of scientists increase dramatically but so did the range of research topics. Rather than attempt to select examples from the almost 9000 papers published by CSHL scientists since 1994, it seems better to give a broad view of the range of research being carried out at the Laboratory in its 125th year.

Bioinformatics and Genomics. Although by the early 1990s the Human Genome Project (HGP) was only just underway it was clear that every area of biological research was going to be changed by the ability to sequence and interpret DNA on a large scale. In 1992 Dick McCombie came to establish a Genome Center at CSHL with the intention of not simply being a major sequencing center, but of ensuring that genomics was integral part of the Laboratory's research programs. The recruitment of Lincoln Stein and Doreen Ware lead the Laboratory into the world of bioinformatics and later quantitative biology. Initially established in the McClintock building, the Genome Center's success demanded a major expansion and the 12-acre campus in Woodbury was purchased in 1998 and renovated by 2001.

The Center rapidly became a key player in the genomics world and was at the forefront in adopting the latest "next-generation" sequencing technologies and developing new techniques. The Center supported the development of Wigler's RDA and ROMA techniques, and later of high-throughput short hairpin RNA libraries for the entire mouse and human genomes that were used to provide screening tools for genetics using cells grown in culture as well as in mice. Contributions to human genetics included the discovery of variations in genome structure in autism by Wigler and in schizophrenia by McCombie. The development in 2007 of the whole-genome exon sequencing (WES) technique by McCombie and Greg Hannon allowed the resequencing of the protein-coding regions of the human genome for disease mutations. In short, after 100 years of genetics, CSHL continues to play a major role in a new style of genetics—whole-genome analysis.

Quantitative Biology. The success of the genomics program led Wigler and Stillman to discussions in 2007 about formalizing and strengthening the growing number of scientists at CSHL who were performing quantitative analysis of genome data. This is an essential part of modern research from genomics, neuroscience, and human diseases to the plant sciences. Appointing faculty with deep expertise in statistics, computer science, mathematics, and physics would provide a rich academic environment for collaboration with wet-lab scientists to advance biological research. With strong support from Jim and Marilyn Simons, the C.V. Starr Foundation, and then CSHL Trustee Landon Clay, an endowment fund was created to support these faculty members. The resulting Simons Center for Quantitative Biology established in 2009 included Gurinder "Mickey" Atwal, Alexander Krasnitz, Dan Levy, Michael Schatz, Ivan Iossifov, Justin Kinney, and the program chair, Adam Siepel. Affiliated faculty already at CSHL included Alex Koulakov and Partha Mitra, who were using theoretical approaches in their studies in neuroscience.

Cancer. Cancer research has a long history at CSHL going back 90 years to the work of Little and MacDowell on cancer in mice in the 1920s and 1930s (Chapter 3). However, it was with the study of tumor viruses and the application of molecular tools beginning in the early 1970s that cancer research at CSHL entered a new phase (Chapter 14). By the early 1980s, cancer research increasingly focused on tumor cells from human or animal cancers, not just those induced by viruses, leading in 1981 to Wigler's isolation of hRAS as the first human tumor–derived oncogene.

Jim and Marilyn Simons.

Scott Lowe, circa 2014.

Alea Mills, circa 2007.

Cancer research at Cold Spring Harbor took off in a new direction beginning in the early 1990s with the start of the international effort to sequence the human genome. It was known that cancer is driven by multiple gene mutations, and developments in sequencing and other technologies made it possible to directly compare the DNA present in tumor cells with the DNA in normal cells, identifying new cancer genes and leading to a better understanding of the genetic basis of cancer. The changes in biochemical pathways caused by mutant genes were investigated with increasing clarity. Mouse models for cancer were developed that more closely mimicked the cancers in patients with respect to treatment outcome, particularly models from Scott Lowe's laboratory that were established in immunocompetent animals and enabled unprecedented investigation of new therapeutic targets. With the discovery of RNA interference by Andy Fire and Craig Mellow in 1998, Greg Hannon's laboratory not only pioneered the understanding of the biochemical mechanisms of RNA interference, but his laboratory in collaboration with Dick McCombie and Steve Elledge (Harvard) also developed short-hairpin RNAs that could inhibit every protein-coding gene in the human or mouse genomes. Hannon and other laboratories, notably those of Scott Lowe and Chris Vakoc, used these RNAs to discover the genes essential for cancer cell proliferation. These genes provided targets for anticancer therapies, some of which are now in clinical trials. Another genetic approach—chromosome engineering—was used by Alea Mills to identify a new tumor-suppressor gene.

Modern cancer research at Cold Spring Harbor is rapidly evolving, with scientists investigating both the adaptive and innate immune systems and their interaction with cancer cells. Others, such as David Spector, David Tuveson, Nick Tonks, and Mikala Egeblad, are growing mouse or human tumor cells in three-dimensional cultures alongside normal cells from the same biopsy, discovering new mechanisms for tumor cell maintenance, and validating new therapeutic targets. Recent expansion of the Cancer Center has included the building of an extension onto the Woodbury research facility to house an advanced preclinical experimental therapeutics research facility, with a focus on assessing combination therapies targeted at cancers. It is clear that the Laboratory's scientists are increasingly working at the interface between fundamental studies on the nature of cancer and the application of those studies in preclinical research. In 2015, an important step in facilitating this work came with the affiliation between Cold Spring Harbor Laboratory and the nearby North Shore Long Island Jewish Health System (page 342).

Rob Martienssen, circa 2010.

Plant Biology. Plant science was the first research program at Cold Spring Harbor and arguably one of the most significant, with George Shull's discovery of hybrid vigor leading to hybrid corn and Barbara McClintock's discovery of controlling elements leading to a new view of how genomes change. The modern-day plant group at Cold Spring Harbor Laboratory studies fundamental mechanisms in plant development and genetics that affect crop productivity, biodiversity, and climate change, which directly relate to the prior discoveries. Rob Martienssen's discovery of the role of RNA interference in the inheritance of heterochromatin provided a molecular explanation for how gene silencing by heterochromatin is inherited and maintained, research recognized as *Science* magazine's Breakthrough of the Year in 2002. Marja Timmerman's work on leaf shape led to a new understanding of how small RNAs control symmetry and leaf angle, an important component of planting density in maize. Zach Lippman's genetic research using tomato has identified a set of genes that explain how hybrids gain improved traits, such as fruit yield, and the nature of the mutation that leads to beefsteak tomatoes. Similarly, David Jackson's finding of a genetic system that controls development of the shoot meristem, through cellular trafficking of transcription factors, has uncovered means to increase yield in maize. Linking plant science and genomics, Dick McCombie has placed Cold Spring Harbor at the forefront of the

Marja Timmerman, circa 2011.

sequencing of plant genomes, including *Arabidopsis*, rice, sorghum, and maize. Annotating these and other genomes by the National Science Foundation's iPlant Collaborative program has enabled Doreen Ware to play a major role in coordinating genome information about plant biology.

Neuroscience. The extent to which conceptual advances in science depend on advances in the development of tools and techniques is not often appreciated. Neuroscientists working in the early 1980s had limited tools but today CSHL neuroscientists have techniques undreamed of 30 years ago. These are used to explore how neural activity and neural circuitry underlie behavior, and how disruptions in these circuits lead to neurological and neuropsychiatric disorders such as Alzheimer's disease, autism, schizophrenia, and depression.

Holly Cline, circa 1996.

The modern-day neuroscience program at CSHL developed with the recruitment of Holly Cline, Robert Malinow, Karel Svoboda, Zach Mainen, and Tony Zador and is constantly evolving. Holly Cline was Director of Research at Cold Spring Harbor from 2002 to 2006 and helped to expand the neuroscience program. In the early 1990s, a talk on a then-new two-photon microscopy technique at a neural imaging course at Cold Spring Harbor, organized by David Tank from Bell Laboratories, led to a major investment in imaging of the brain. Karel Svoboda was recruited to Cold Spring Harbor in 1997 from Bell Labs where he had worked with Winfried Denk and David Tank. Denk, Svoboda, and Tank pioneered two-photon microscopy and its application to brain science, and in 2015, they received the prestigious Brain Prize. Svoboda and Robert Malinow developed techniques to study neural activity and structural changes in the brain of living animals. However, the success of the brain imaging program at Cold Spring Harbor caused the new Howard Hughes Medical Institute (HHMI) Janelia Farm Research Center in northern Virginia to recruit three Cold Spring Harbor scientists to start its new neuroscience research initiative that would integrate brain science and imaging.

Karel Svoboda, circa 2014.

Thus, in 2006, Bruce Stillman decided to take the neuroscience program in an entirely different direction, based on two separate advances. In 2004 Mike Wigler had discovered a very high frequency of copy number variations (CNVs) in the DNA of specific segments of the human genome and began to study these variations in autism. By 2007, he had showed a 10-fold increase in CNVs in people with autism compared to people without autism, and he later showed that, combined with specific point mutations, ~60% of autism could be explained by de novo or inherited mutations. This research was almost entirely supported by the Simons Foundation, a major funder of autism research throughout the world. The research also led to Dick McCombie and Shane MacCarthy studying CNVs in schizophrenia and bipolar disorders with major funding from the Stanley Medical Research Institute.

Tony Zador, circa 2014.

Beginning in 2006, in parallel with these genetic discoveries on cognitive disorders, Zach Mainen, Adam Kepecs, and Tony Zador, and more recently Anne Churchland, developed techniques using olfactory and auditory cues to study cognition in rats and mice. These techniques enabled studies on decision-making, attention, confidence, and working memory that had previously been restricted to awake behaving monkeys. The rodent cognition program was thus born and represents a major initiative in the current neuroscience program. These studies have been complemented by the study of brain development and circuits, led by Josh Huang, Partha Mitra, and Tony Zador, including the use of DNA-based bar coding to map the connections between synapses in the brain.

The Double Helix Medals

Anne Churchland, circa 2013.

The Cold Spring Harbor Laboratory Association continues to be a major supporter of the Laboratory providing essential unrestricted funding for research. Its fund-raising activities expanded to include an annual golf tournament at Piping Rock Club and, most notably, the Double Helix Medals Awards Dinner. Inspired by Cathy Soref, each year outstanding individuals are acknowledged for their contributions to science, philanthropy, and raising public awareness of science or health disorders. The first ceremony was held in 2006, when the recipients were Muhammad Ali for his public campaign in support Parkinson's disease research; Suzanne and Bob Wright for their work in bringing attention to autism; and Nobel laureate and former CSHL scientist Phillip Sharp for his lifelong contributions to biomedical research and the biotechnology industry. Many individuals contribute to the success of the awards ceremonies, not least Phil Donahue and Deborah Norville, who have acted as emcee.

Recipients of the Double Helix Medal

Year	Recipients	Year	Recipients
2006	Muhammad Ali Phillip Sharp Suzanne and Bob Wright	**2007**	Richard Axel David Koch Michael Wigler
2008	Sherry Lansing Marilyn and James Simons J. Craig Venter James D. Watson	**2009**	Herbert W. Boyer Stanley N. Cohen Kathryn W. Davis Maurice "Hank" Greenberg
2010	Mary-Claire King Evelyn Lauder John Nash	**2011**	Kareem Abdul-Jabbar Temple Grandin Harold Varmus
2012	Michael J. Fox Art Levinson Mary Lindsay	**2013**	Peter Neufeld Robin Roberts Barry Scheck
2014	Matthew Meselson Andrew Solomon Marlo Thomas	**2015**	David Botstein Katie Couric Anne Wojcicki

Bob Wright, Lonnie Ali, Muhammad Ali, Suzanne Wright, and Bruce Stillman (2006).

John Nash, Mary-Claire King, and Evelyn Lauder (2010).

Harold Varmus, Kareem Abdul-Jabbar, and Temple Grandin (2011).

Michael J. Fox, Tracy Pollan, Mary Lindsay, Art Levinson, Jamie Nichols, and Bruce Stillman (2012).

Peter Neufeld, Robin Roberts, Barry Scheck, and Bruce Stillman (2013).

Bruce Stillman, Matt Meselson, Marlo Thomas, and Andrew Solomon (2014).

The Double Helix Awards.

The Hillside Campus

The success of the neuroscience program and the increasing use of genomics and bioinformatics in all areas of biology required additional research space so that these fields could contain a critical mass of researchers at Cold Spring Harbor. At the same time, it has always been apparent that the Laboratory needed more and better accommodations for the many visitors who attend its meetings and courses. However, with the rezoning of the Laboratory's land in the Village of Laurel Hollow, any future expansion needed to be strategically placed to maximize the available space and also to be developed in keeping with the beauty of the main campus. Watson and Stillman therefore sought funds from Charles and Helen Dolan so that they could commission Centerbrook Architects to develop a master plan to use a relatively level area at the top of the main campus as well as the hillside above Urey Cottage.

The plan that emerged proposed that the upper area would become a quadrangle with a lawn framed by accommodation for students and

The conceptual design for the Hillside campus. Dolan Hall and the Beckman Laboratory are at the lower left.

visitors, offices and seminar rooms for the WSBS, an archive building, a library, and a great hall. Six laboratories would be built on the hillside dropping down to Urey Cottage, each laboratory to look like a small building so that they would not look like a monolithic structure from across the harbor. It was a spectacular project but just as the conceptual designs were completed in October 2001 the United States entered a recession.

Fortunately, the master plan was intended to be implemented in stages, and in 2005, following extensive consultations with the Village of Laurel Hollow and the Laboratory's neighbors, ground was broken on the first stage, the construction of six laboratories. These were named in honor of substantial gifts from Donald Everett Axinn, Nancy and Frederick DeMatteis, David H. Koch, William L. and Marjorie A. Matheson, Leslie and Jean Quick, and the Wendt Family. On July 22, 2008, guests, elected officials, and staff gathered for the topping out ceremony when a stainless steel pyramid was lifted to the top of the Laurie and Leo Guthart Tower. The laboratories were occupied during 2009

Topping out ceremony.

The Hillside campus from across the Harbor. The photograph spans 114 years from the Jones Laboratory (1895), lower left, *to the Hillside Laboratories (2009).*

and enabled a major recruitment drive, becoming home to a new wave of young scientists. Of the 19 faculty members occupying laboratories on the Hillside campus, 15 have been recruited since the buildings were opened.

Creating Therapeutic Drugs

A phrase increasingly heard is "translational research"—that is, research that is directed toward "translating" the findings of basic research into practical applications. This is hardly a modern idea, as scientists have always wanted to see their work put to good use. Since its beginning, scientists at Cold Spring Harbor were practicing translational research: George Shull's hybrid corn led to a revolution in American agriculture, Willis Swingle's work isolating hormones from the adrenal cortex led to a treatment for Addison's disease, and during World War II Demerec used his knowledge of bacterial genetics and mutations to make strains of *Penicillium* that overproduced penicillin. At present, CSHL scientists are enthusiastically looking to put their expertise and knowledge to practical use.

Chris Vakoc works on acute myeloid leukemia (AML) and, using the RNAi techniques developed in Greg Hannon's laboratory and the mouse models developed by Scott Lowe, identified a protein Brd4 that helps regulate gene expression, particularly the *c-Myc* gene, one of the most common oncogenes in human cancers. Vakoc and Lowe, with collaborators at Dana-Farber Cancer Center, found that inhibitors of Brd4 extended the survival of leukemia-bearing mice, and in 2013 these inhibitors entered Phase I clinical trials in leukemia patients. Early reports are promising, suggesting clinical responses in a subset of treated patients.

It takes a long time to understand the molecular machinery underlying life processes and to translate that knowledge into a drug. In the early 1990s, Nick Tonks discovered that enzymes called tyrosine phosphatases

Chris Vakoc, circa 2013.

play a key role in signal transduction, the process by which cells respond to stimuli in their environment. The function of these enzymes is frequently disrupted in human diseases, including diabetes and cancer, and the Tonks laboratory, using knowledge gained from very basic studies on how these enzymes function, has been working to develop drugs that modulate their activity. It has not been easy but using their knowledge of how the enzymes work at the atomic level, a small molecule has been developed that will be entering Phase I clinical trials in patients with metastatic breast cancer.

Cold Spring Harbor Laboratory's efforts in developing therapies are not restricted to cancer, and Adrian Krainer has worked on spinal muscular atrophy (SMA) for many years. SMA is a debilitating neuromuscular disorder caused by mutations in the gene *SMN1*. In 1999 a second gene, *SMN2*, was found but this has a defect in splicing of RNA so it produces only small amounts of functional SMN protein. The reduced level of SMN protein leads to the disorder. RNA splicing was discovered at Cold Spring Harbor and MIT in 1977, and its biochemical steps worked out at Harvard and MIT in 1984. In 1986 Krainer came from Harvard to Cold Spring Harbor as the first CSHL Fellow, working on the mechanism and regulation of splicing. Using his intimate knowledge of the splicing machinery, Krainer was able to correct the splicing defect in the *SMN2* RNA so that it produces enough protein to allow motorneuron survival. In collaboration with Isis Pharmaceuticals, an "antisense" oligonucleotide drug, ISIS-SMNRx, was developed, which successfully completed Phase I and Phase II trials. There is great hope that it will prove successful in ongoing Phase III trials and then be approved by the Federal Drug Administration to provide the first treatment for this devastating childhood genetic disorder.

Adrian Krainer, circa 2010.

The work of Vakoc, Tonks, and Krainer is a harbinger of the next phase in the Laboratory's development, ensuring that CSHL scientists can carry the fruits of their research beyond the walls of their laboratories.

David Tuveson, circa 2013.

A first formal step to promote translational research came in 2013 when David Tuveson was recruited to lead a new CSHL Cancer Therapeutics Initiative. This has three components: identifying therapeutic targets using techniques developed at Cold Spring Harbor, using mouse models to understand the molecular details of these targets, and using these mouse models to evaluate the efficacy of drugs that might be used in patients. The new Preclinical Experimental Therapeutics Facility at the Woodbury Genome Center is an important component of this initiative. This facility will house a number of cancer diagnostic and therapeutic resources in support of the Cancer Therapeutics Initiative.

The nature of modern biomedical science has been such a success that institutions such as Cold Spring Harbor Laboratory are often finding that their faculty have opportunities to translate their basic science discoveries into practical application. However, this comes at a significant cost, drawing critical endowment funds away from supporting basic, discovery science. Although the CSHL endowment had risen to $450 million by 2015, it was still not sufficient to support the Laboratory's infrastructure and basic research, as well as support expensive translations opportunities, creating a dilemma for Bruce Stillman. Furthermore, as such opportunities arise, it is increasingly necessary for Cold Spring Harbor scientists to have close collaborations with clinicians to guide the research.

Therefore, a second and very significant development came in 2015 when Bruce Stillman and Michael Dowling, President and Chief Executive Officer of the North Shore–LIJ Health System, signed an agreement establishing an affiliation between the institutions to focus on cancer. Projected for an initial 10 years with an investment of more than $120 million, the affiliation will promote the integration of research scientists, clinical translational researchers, and cancer clinicians at CSHL and North Shore–LIJ. With the recruitment of clinician scientists who will hold joint appointments between the institutions, the collaborative

On April 2, 2015, Bruce Stillman and Michael Dowling, President and Chief Executive Officer of the North Shore–LIJ Health System, signed an agreement establishing an affiliation between the institutions.

preclinical research will be the basis of advanced-phase clinical trials to be conducted both at North Shore–LIJ facilities and collaborating outside medical centers. Patients will benefit from increased access to these innovative clinical studies, and CSHL scientists will benefit from a more rapid application of their promising findings. The funds provided by this initiative for the translational research in cancer have enabled the endowment funds that are so critical for the future success of Cold Spring Harbor Laboratory to be employed for basic, discovery science. It is this science that has the potential to benefit mankind in ways that are unimaginable today.

The Future

As Cold Spring Harbor enters its 125th year, it finds itself much stronger and better able to face the future than ever before. Its research programs have expanded and reached a critical mass, its scientists are pursuing exciting research that will help us all, and the renown of its education programs at all levels continues to grow. None of this would be possible without the philanthropy of individuals and foundations, and it may not be realized the extent to which philanthropy has played a role in creating and sustaining the community of science at Cold Spring Harbor.

The lands for original institutions by the harbor—the Biological Laboratory and the Station for Experimental Evolution—were given by John D. Jones and the Wawepex Society. The institutions were established by the Brooklyn Institute of Arts and Sciences and by Andrew Carnegie's legacy, the Carnegie Institute of Washington, bodies whose goal was not to make money but to increase learning and knowledge. In the early 1920s it was the philanthropy from the Biological Laboratory's neighbors and the creation of the Long Island Biological Association (LIBA) that saw it through the financial crisis following its abandonment by the Brooklyn Institute. Throughout the 1920s and 1930s, LIBA and Carnegie were essentially the only sources of funding for research and teaching, while The Rockefeller Foundation also supported meetings during the 1930s, playing a major role in sustaining the Biological Laboratory, and providing funds for the construction of new buildings. Friends of the Laboratory also made significant contributions to the Cold Spring Harbor landscape in this period (e.g., Blackford Hall and the Nichols Building).

It was not until after the Second World War that the federal government began funding scientific research. President Roosevelt asked Vannevar Bush, Director of the Carnegie Institute of Washington but then acting as Director of the Office of Scientific Research and Development, to recommend how the high level of research that had gone on in the war could be sustained in peacetime. Bush's report, *Science: The Endless Frontier,* recommended that investment by the federal government in research would help pay off the considerable national debt caused by the war expenditures. As a result, it ushered in the modern era of research in the United States and a major economic expansion during the 1950s and

1960s. In 1946, the Public Health Service created a program of extramural research grants and fellowship awards, but the great expansion in biomedical funding did not come until the early 1960s leading to the current dependence of research funding on the National Science Foundation and National Institutes of Health.

In the meantime, the Laboratory's friends continued to provide support culminating, in the early 1970s, with the gifts from the Robertson family (Chapter 14). The gift of $8.5 million dollars formed the bedrock of the Laboratory's quest for endowment, and without the endowment Robertson provided for the Banbury Center, it would have been hard for CSHL to sustain the Center and its activities. More recently, Jim and Marilyn Simons donated $50 million as an endowment for the Quantitative Biology program and considerable other funds to support research programs at Cold Spring Harbor, including a major effort in autism research.

The importance of the CSHL endowment has grown over the years, especially over the past decade or so when the NIH budget has fallen by ~25% in real terms. The income from the endowment has spared CSHL some of the financial stresses other independent research institutes have experienced. For example, the Marine Biological Laboratory at Woods Hole in Massachusetts, established at roughly the same time as Cold Spring Harbor Laboratory, has had to affiliate with the University of Chicago with a consequent loss of independence. Although the current endowment of ~$450 million is not sufficient to sustain CSHL going forward, it has buffered the Laboratory from the worst of financial storms and kept the Laboratory at the forefront of research and education.

What, then, of the future of Cold Spring Harbor Laboratory? The pace of progress in science is unprecedented, particularly in the field of genetics, the core strength of the institution for nearly its entire history, and genomics. The story of CSHL over these years has been one of growth that has responded to the changing times, sometimes in spurts

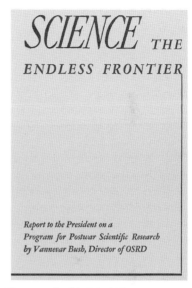

Vannevar Bush's July 1945 report, commissioned by President Roosevelt, on recommendations for promoting postwar American scientific research.

In this sculpture by Pablo Eduardo, the young Charles Darwin, with a Galapagos finch perched on his staff, gazes across Cold Spring Harbor.

of rapid expansion and at other times through steady expansion of both science and education. There is every reason to expect that the next *A Short History of Cold Spring Harbor Laboratory* will have to be long!

WIGLER AND COLLEAGUES EXPLORE THE GENOME

Summary

In 2004, Michael Wigler and colleagues discovered that a type of structural variation in the human genome—large-scale copy-number polymorphism—is common and serves to distinguish one individual from another. The discovery has profound implications for the origins of human diversity and evolution, and lies at the root of many common genetic diseases, from cancer to autism.

The Research

Cancers differ from normal cells by a process of mutation. Finding these mutations has motivated Michael Wigler throughout his career in research. Arriving at CSHL in 1978, he began applying techniques he had developed at Columbia University in the laboratory of future Nobel laureate Richard Axel. He and his CSHL lab developed a set of increasingly powerful tools for comparing genomes, searching for minor differences that might have large phenotypic effects. Among these methods was DNA cotransformation, a means of engineering cells that Wigler, Axel, and Saul Silverstein had developed. It enabled Wigler's codiscovery in 1981–1982 of members of the RAS family of oncogenes and their role in cancer.

In the 1990s, Wigler, Nikolai Lisitsyn, and Robert Lucito devised powerful methods for subtractive or differential DNA hybridization. Among these were representational difference analysis (RDA), a subtractive hybridization technique that brought to the fore differences between paired normal and tumor genomes. In 1997, the PTEN tumor suppressor gene was discovered by the lab, using RDA to study 12 primary breast tumors. A variant method called representational oligonucleotide microarray analysis (ROMA) was an array-based hybridization method, and for a time, it was the most powerful means of searching for structural variations in cancer. Both RDA and ROMA depended on a conceptual advance, genome "representation," based on pseudo-random reproducible samplings of the entire the genome. The group achieved efficient hybridization to arrays of probes, thereby allowing the direct and accurate measure of copy number at unprecedented resolution. The typical copy number for a segment in the diploid genome is two, but as cancers duplicate or lose genome segments, copy numbers can vary widely. Mathematical methods were needed to interpret the data, as well as implementation as computer code. Developing a team to do this became the seed for the Simons Center for Quantitative Biology at CSHL.

Using ROMA in 2003–2004, Wigler's lab compared cancer samples from various patients to the normal DNA of a single unrelated healthy person. They noticed that most

(Left to right) *Jonathan Sebat, Michael Wigler, and Lakshmi Muthuswamy in 2007.*

Continued

cancers carried the same identical copy-number events when compared to the control, but *not* when compared to the individual patients from whom the cancer samples were derived. Incidentally, then, they had discovered structural variation between normal genomes.

The Significance

In a landmark 2004 *Science* paper, Wigler, Lakshmi Muthuswamy, Jonathan Sebat, and others reported finding nearly 100 variations among 20 persons, many of which contained genes. Because the dosage of such genes was likely different between such people, it was immediately clear that copy-number variation would be a rich source of human variation. It was also clear that most variation was rare and therefore likely of recent origin. As such, they expected to find, and did subsequently, a relatively high rate of spontaneous (de novo) germline generation of such variants. These were discovered by the now classic family-trio approach, comparing a child to its parents. The lab quickly used this approach to study autism, leading to a new theory published in 2007 about its origins and the discovery of many causative genes. They further predicted and later showed that germline events could contribute to other known pediatric disorders such as congenital heart disease and cancer. Methods for comparing the sequence of a child's genome to that of the parents have led to an appreciation of the role of rare and new mutations in human genetic disorders.

The Scientist

Michael H. Wigler was born in New York on September 3, 1947, and grew up in Garden City. After earning a B.A. at Princeton University, he began medical studies at Rutgers in 1972, transferring the following year to Columbia University to study microbiology. He was awarded the Ph.D. degree in 1978. He has been at Cold Spring Harbor Laboratory since 1978, serving as a professor of mammalian cell genetics and, since 1986, American Cancer Society Lifetime Research Professor. He has been elected to the National Academy of Sciences and the American Academy of Arts and Sciences and has won many awards and honors including CSHL's Double Helix Award. He holds 30 U.S. patents and has founded and been active in a number of biotechnology companies.

— *P.T.*

GREGORY HANNON AND MECHANISM OF RNAi

Summary

RNA interference (RNAi) is a way of altering gene expression in living cells that has brought about a revolution in biological research. Greg Hannon has not only helped define how RNAi works, but has also created tools that allow researchers across the world to harness its power.

The Discovery

Researchers have long exploited model systems that make genetic manipulations naturally easy, like yeast, worms, and flies. But the same experiments in mammalian cells and organisms were stymied by technological limitations.

All that changed with the discovery of RNA interference (RNAi), a biological process that cells use to control gene expression. Hannon set out to define how RNAi works. In a paper published in *Nature* in 2000, his team described how they isolated the RISC protein complex, a nuclease that uses short RNAs to degrade specific mRNA transcripts, silencing the gene. Hannon and others went on to dissect the RISC complex, identifying its molecular components. He found the enzyme Dicer, a protein that converts long double-stranded RNA into smaller fragments that guide the silencing process. In addition, Hannon identified Argonaute, the so-called Slicer nuclease at the core of RISC that degrades target mRNAs, while Leemor Joshua-Tor, Hannon's colleague at CSHL, used X-ray crystallography to determine the atomic structure of the RISC complex.

Hannon collaborated with Steve Elledge and Scott Lowe to use shRNAs in animals and to create shRNA libraries that are capable of silencing nearly every gene in the human, mouse, and rat genomes.

Significance

Around the world, researchers began to use RNAi to investigate the pathways that govern normal cellular growth and organism development. Scientists are also employing the

Greg Hannon in his lab, circa 2007.

technology to search for genetic weaknesses in diseases as varied as cancer, infertility, and mental illness. Already, the technology has led to success. For example, researchers at CSHL recently used RNAi to discover a protein target for acute myeloid leukemia, and an inhibitory drug against the protein is now in clinical trials.

The Scientist

Greg Hannon was an undergraduate at Case Western University, and he remained at Case Western for his graduate work. In 1992, Hannon came to CSHL as a postdoctoral fellow in David Beach's laboratory to work on the cell cycle, but soon focused on RNAi. In 2005, he became an investigator with the Howard Hughes Medical Institute and was later named a member of the National Academy of Sciences. Hannon has since relocated his laboratory to Cancer Research UK Cambridge Institute.

— *J.J.*

ROBERT MARTIENSSEN, TRANSPOSONS, AND EPIGENETIC CONTROL OF GENE EXPRESSION

Summary

In 2002–2004, plant geneticist Robert Martienssen and colleagues used microarrays and other genome-age technology to discover the role that transposons play in heterochromatin and in the epigenetic control of genes, in the process providing an example of transposon function in development and suggesting their impact in evolution—ideas advanced by Barbara McClintock at Cold Spring Harbor Laboratory decades before.

The Research

In 1988, Robert Martienssen gave his first talk on plant genetics at Cold Spring Harbor to an audience of six people. Three were Nobel laureates, and one, Barbara McClintock, not coincidentally was his intellectual hero. It was she who had discovered transposable elements (TEs)—segments of DNA that could jump at random within the genome—and who had first ventured that these genetic oddities were not merely disrupters but were in fact important agents in the development and evolution of organisms. From his earliest research projects, in some of which TEs were used as tools for genome exploration, Martienssen was clear about the real goal, which was not instrumental but foundational: "to show the ways in which transposons control gene expression," as McClintock had begun to demonstrate in her work on maize, albeit through a scientific glass, darkly, half a century before Martienssen arrived at Cold Spring Harbor.

In 2000, completion of the project to sequence the first plant genome—that of the mustard plant *Arabidopsis thaliana*—suggested Martienssen's next steps. Two isolated regions of heterochromatin called "knobs" offered the first contiguous sequence of heterochromatin in a higher eukaryote, thus "allowing silencing and methylation to be tested through microarrays." Heterochromatin was known to be composed of highly compacted, largely "silent," repetitive DNA. It was regarded inert and uninteresting. That picture changed markedly with research published by Martienssen and colleagues in 2002, recognized by *Science* as the "breakthrough of the year." It was work in which the nature of heterochromatin and associated gene silencing became bound up in the then new world of small RNA biology. In fission yeast mutants lacking components of the RNA interference (RNAi) machinery, they observed defects in heterochromatin formation at centromeres. In yeast cells lacking small interfering RNAs (siRNAs), chromatin dynamics became unbalanced, impairing cell division. The team had discovered that heterochromatic silencing depended on the processing of repeat RNA transcripts into siRNAs, which then directed chromatin modification. In 2004, Martienssen and colleagues, publishing

Robert Martienssen and Barbara McClintock examining experimental maize plants in 1989.

Continued

in *Nature*, showed that heterochromatin in the knob regions of *A. thaliana* was defined by TEs and associated tandem repeats. Using microarrays of their own design, they demonstrated how, in an *A. thaliana* mutant lacking the *DDM1* gene—which encodes an enzyme that modifies chromatin via methylation—silencing marks were lost. The result was massive TE activation. They concluded that TEs silenced genes epigenetically, and not only in heterochromatin. TEs "scattered throughout the genome," they ventured, had the potential to "regulate development," but exerted such control only when they integrated within or very near genes they acted upon.

Significance

Martienssen's team had made fundamental discoveries about the nature of heterochromatin, as well as about the mechanisms that silence it, that would eventually be observed in most species, including mammals. They had also vividly demonstrated how TEs can control gene expression genome-wide. They offered the example, in *A. thaliana*, of the *FWA* gene, situated in a euchromatic region. It was repressed by a TE lodged in its promoter.

Silencing was due to methylation of the promoter, in the absence of which the gene was exposed to the transcriptional machinery and expressed. Loss of methylation-associated repression caused late flowering and ultimately sterility. In this way, the TE's insertion, a random event in the plant's evolutionary history, had a lasting impact on developmental regulation in *A. thaliana*; the mechanism through which it exerted this influence was epigenetic.

The Scientist

Robert Martienssen was educated at Cambridge University, earning his B.A. in Genetics and his Ph.D. at the Plant Breeding Institute. After postdoctoral research at the University of California, Berkeley, he came to Cold Spring Harbor in 1989, rising to a full professorship in 1996. In 2012, he was named an Investigator of the Howard Hughes Medical Institute–Gordon and Berry Moore Foundation. He is a Fellow of the Royal Society (U.K.) in addition to many other wards and honors. He holds several U.S. patents arising from the invention of techniques used in plant genetics.

— *P.T.*

Chairmen of CSHL Board of Trustees and Presidents of LIBA/CSHLA

Chairmen of the Cold Spring Harbor Laboratory Board of Trustees (1963 to present):

1963–1967	Edward L. Tatum
1967–1973	H. Bentley Glass
1974	Robert H. P. Olney
1975–1979	Harry Eagle
1980–1985	Walter H. Page
1986–1991	Bayard D. Clarkson
1992–1997	David L. Luke III
1998–2003	William R. Miller
2004–2010	Eduardo G. Mestre
2011–present	Jamie C. Nicholls

Presidents of the Long Island Biological Association (1926–1991) and the Cold Spring Harbor Laboratory Association (1992 to present):

1924–1926	Timothy S. Williams	1998	Vernon Merrill
1926	Walter B. James	1998–2000	James Spingarn
1927–1940	Arthur Page	2000–2002	David H. Deming
1940–1952	Robert Cushman Murphy	2002–2004	Trudy H. Calabrese
1952–1957	Amyas Ames	2004–2006	Joseph T. Donohue
1957–1972	Walter H. Page	2006–2008	Pien Bosch
1972–1985	W. Edward Pulling	2008–2010	Timothy S. Broadbent
1985–1992	George W. Cutting	2010–2013	Sandy Tytel
1993–1994	Mary D. Lindsay	2014–present	Frank O'Keefe
1995–1997	John P. Cleary		

Further Reading

Cairns J, Stent GS, Watson JD. 2007. *Phage and the origins of molecular biology: The centennial edition.* Cold Spring Harbor Laboratory Press, Cold Spring Harbor, NY.

Essays on Max Delbrück but including many references to Cold Spring Harbor of the 1950s.

Comfort NC. 2001. *The tangled field: Barbara McClintock's search for the patterns of genetic control.* Harvard University Press, Cambridge, MA.

The definitive study of Barbara McClintock.

Cook-Deegan R. 1994. *Gene wars: Science, politics, and the human genome.* WW. Norton, New York.

Covers the origins of the Human Genome Project and Watson's departure.

Earle, WK. 1966. *Out of the wilderness. Being an account of aspects of the settlement of Cold Spring Harbor, Long Island, and the activities of some of the settlers, from the beginning to the Civil War.* Whaling Museum Society, Inc., Cold Spring Harbor, NY.

Cold Spring Harbor in the days before the Biological Laboratory.

Fischer EP, Lipson C. 1988. *Thinking about science: Max Delbrück and the origins of molecular biology.* WW. Norton, New York.

A biography of Max Delbrück and his work in the life sciences.

Hiltzik LR. 1993. "*The Brooklyn Institute of Arts and Sciences' Biological Laboratory, 1890–1924: A history.*" PhD thesis, State University of New York, Stony Brook.

An unpublished, detailed study of the origins of the Biological Laboratory.

Hughes RC. 2014. *Cold Spring Harbor (Images of America).* Arcadia, Charleston, SC. Includes many photographs showing life at Cold Spring Harbor at the time of the founding of the Biological Laboratory and Station for Experimental Evolution.

Inglis JR, Sambrook J, Witkowski JA. 2003. *Inspiring science: Jim Watson and the age of DNA.* Cold Spring Harbor Laboratory Press, Cold Spring Harbor, NY.

Essays by Watson's colleagues at Harvard and Cold Spring Harbor Laboratory.

Judson HF. 2013. *The eighth day of creation: Makers of the revolution in biology, Commemorative edition.* Cold Spring Harbor Laboratory Press, Cold Spring Harbor, NY.

The classic history of the development of molecular genetics from the 1930s to the 1960s.

Kevles DJ. 1985. *In the name of eugenics: Genetics and the uses of human heredity.* Alfred A. Knopf, New York.

A comprehensive study of eugenics.

Kohler, RE. 2002. *Landscapes and labscapes: Exploring the lab-field border in biology.* University of Chicago Press, Chicago.

Includes a short but interesting discussion of Banta's work in the field (collecting in caves) and in the laboratory (his artificial "cave" at Cold Spring Harbor).

McElheny VK. 2003. *Watson and DNA: Making a scientific revolution.* Perseus, Cambridge, MA.

A biography of James D. Watson.

Rettig RA. 1977. *Cancer crusade: The story of the National Cancer Act of 1971.* Princeton University Press, Princeton, NJ.

An account of the politics and machinations behind the National Cancer Act.

Stahl FW. 2000. *We can sleep later: Alfred D. Hershey and the origins of molecular biology.* Cold Spring Harbor Laboratory Press, Cold Spring Harbor, NY.

Essays honoring Alfred Hershey.

Wallace B. n.d. *On the fringe of glory: The Cold Spring Harbor Laboratories of the 1950s.* Unpublished manuscript.

A personal account of life at Cold Spring Harbor Laboratory in the 1950s.

Watson EL. 1991. *Houses for science: A pictorial history of Cold Spring Harbor Laboratory.* Cold Spring Harbor Laboratory Press, Cold Spring Harbor, NY.

A beautiful book describing the history of CSHL through its buildings.

Watson EL. 2008. *Grounds for knowledge: A guide to Cold Spring Harbor's landscapes and buildings.* Cold Spring Harbor Laboratory Press, Cold Spring Harbor, NY.

A survey of the landscape of the campus which is so important to the character of CSHL.

Watson JD. 2001. *A passion for DNA: Genes, genomes, and society.* Cold Spring Harbor Laboratory Press, Cold Spring Harbor, NY.

A collection of James D. Watson's essays including some drawn from the CSHL Annual Reports.

Witkowski JA. 2000. *Illuminating life: Selected papers from Cold Spring Harbor, Volume 1*. Cold Spring Harbor Laboratory Press, Cold Spring Harbor, NY.

A selection of notable publications from Cold Spring Harbor, from 1903 to 1969 with introductory essays.

Witkowski JA, Gann A, Sambrook JF. 2008. *Life illuminated: Selected papers from Cold Spring Harbor, Volume 2*. Cold Spring Harbor Laboratory Press, Cold Spring Harbor, NY.

A further collection of CSHL papers from 1972 to 1994, with essays by the scientists who carried out the research.

Witkowski JA, Inglis JR. 2008. *Davenport's dream: 21st century reflections on heredity and eugenics*. Cold Spring Harbor Laboratory Press, Cold Spring Harbor, NY.

A reprinting of Davenport's *Heredity in relation to eugenics* with introductory essays.

Witkowski JA. 2010. *75 Years in science at the Cold Spring Harbor Laboratory Symposia on Quantitative Biology*. Cold Spring Harbor Laboratory Press, Cold Spring Harbor, NY.

A booklet celebrating the Symposia with essays by notable scientists.

Biographies of many of the scientists mentioned in this book have appeared in the series Biographical Memoirs published by the National Academy of Sciences (U.S.):

Blakeslee, Albert Francis (1874–1954), by Sinnott EW. 1959. *Biogr Mem Natl Acad Sci* **33:** 1–38.

Bridges, Calvin Blackman (1889–1938), by Morgan TH. 1940. *Biogr Mem Natl Acad Sci* **22:** 31–48.

Bush, Vannevar (1890–1974), by Wiesner JB. 1979. *Biogr Mem Natl Acad Sci* **50:** 89–117.

Castle, William Ernest (1867–1962), by Dunn LC. 1965. *Biogr Mem Natl Acad Sci* **38:** 33–80.

Demerec, Milislav (1895–1966), by Glass B. 1971. *Biogr Mem Natl Acad Sci* **42:** 1–27.

Goldschmidt, Richard Benedict (1878–1958), by Stern C. 1967. *Biogr Mem Natl Acad Sci* **39:** 141–192.

Kaufmann, Berwind Petersen (1897–1975), by Lewis E. 2004. *Biogr Mem Natl Acad Sci* **85**: 125–135.

Little, Clarence Cook (1888–1971), by Snell GD. 1975. *Biogr Mem Natl Acad Sci* **46**: 241–263.

Riddle, Oscar (1877–1968), by Corner GW. 1974. *Biogr Mem Natl Acad Sci* **45**: 427–465.

Tatum, Edward Lawrie (1909–1975), by Lederberg J. 1990. *Biogr Mem Natl Acad Sci* **59**: 357–386.

Other useful sources of biographical information include:

Crow JF. 2001. Plant breeding giants: Burbank, the artist; Vavilov, the scientist. *Genetics* **158**: 1391–1395.

Crow JF. 2001. Shannon's brief foray into genetics. *Genetics* **159**: 915–917.

Hilliker A. 2012. Arthur Chovnick (1927–2011) geneticist. *Genetics* **190**: 827.

MacDowell EC. 1946. Charles Benedict Davenport 1866–1944. A study in conflicting influences. *BIOS* **17**: 3–50.

Witkin EM. 2002. Chances and choices: Cold Spring Harbor 1944–1955. *Ann Rev Microbiol* **56**: 1–15.

Web Resources

http://cshl.edu

Cold Spring Harbor Laboratory's main website.

http://library.cshl.edu/archives

The Cold Spring Harbor Laboratory Library and Archives houses a rich repository of rare books, manuscripts, photographs, and scientific reprints documenting genetic research and the work of the Laboratory faculty since 1890.

http://www.eugenicsarchive.org/eugenics/

The DNA Learning Center's "Image Archive on the American Eugenics Movement" is the primary online source for materials relating to eugenics. It includes photographs of leading eugenicists as well as images of articles, documents, charts, and pedigrees. The materials are drawn from around the world, although focusing on the Eugenics Records Office.

Illustration Sources

All images and illustrations were provided courtesy of Cold Spring Harbor Laboratory Library and Archives, with the exception of the following, which also include those items from the Special Collections of the Archives. Every effort has been made to contact the copyright holders of figures used in this text. Any copyright holders we have been unable to reach, or for whom inaccurate information has been provided, are invited to contact Cold Spring Harbor Laboratory Press.

Chapter 1: **p. 1, top,** From Morgan TH. 1934. *Embryology and genetics.* Columbia University Press, New York; **p. 1, bottom,** Wikipedia (https://en.wikipedia.org/wiki/StazioneZoologica#/media/File: Stazione_zoologica_Dohrn.jpg); **p. 2, top,** from Lattin FH. 1895. *Penikese: A reminiscence.* Frank H. Lattin Publisher, Albion, NY; **p. 2, bottom,** Brooklyn Museum Archives, Lantern Slide Collection; **p. 4, top,** from Bean TH. 1903. *The food and game fishes of New York: Notes on their common names, distribution, habits, and mode of capture.* J.B. Lyon Company, Albany, NY; **p. 4, bottom,** Collection of the New-York Historical Society; **p. 5, bottom,** Wikipedia (http://en.wikipedia.org/wiki/Herbert_William_Conn); **p. 9, top,** *The New York Times,* September 9, 1892; **p. 11, right,** From Davenport CB. 1903. *The animal ecology of the Cold Spring sand spit, with remarks on the theory of adaptation,* Vol. X: pp. 157–176. Decennial Publications U. Chicago, Chicago.

Chapter 2: **p. 13,** From Davenport CB. 1936. *Statistical methods in biology, medicine and psychology.* John Wiley & Sons, Inc., New York; **p. 16,** from Carnegie Institution of Washington, Year Book No. 1, 1902, published by the Institution, Washington, U.S.A., January, 1903; **p. 21,** from Carnegie Institution of Washington, Year Book No. 3, 1904, published by the Institution, Washington, U.S.A., January 1905; and Year Book No. 4, 1905, published by the Institution, Washington, U.S.A., January 1906; **p. 25,** from Shull G. 1911. *The American Naturalist* 45: 234–252, published by The University of Chicago Press for The American Society of Naturalists; **p. 26,** courtesy of Purdue University Libraries Archives and Special Collections; **p. 27, top,** courtesy of The American Museum of Natural History, Image 314972; **p. 27, bottom,** from

Morgan TH. 1919. *The physical basis of heredity.* J.B. Lippincott, New York; **p. 29,** from Carnegie Institution of Washington, Year Book No. 5, 1906, published by the Institution, Washington, U.S.A., January, 1907.

Chapter 3: **p. 34,** Courtesy of Jan Witkowski; **p. 35,** Wikipedia (http://en.wikipedia.org/wiki/File:Edward_Henry_Harriman_1899. jpg); **p. 36,** courtesy of Harry H. Laughlin Papers, Pickler Memorial Library Collection, Truman State University; **p. 38, p. 39,** digital images courtesy of Susan Lauter, Dolan DNA Learning Center; **p. 40, top,** digital images courtesy of Susan Lauter, Dolan DNA Learning Center; **p. 40, bottom,** courtesy of American Philosophical Society; **p. 42,** courtesy of Harry H. Laughlin Papers, Pickler Memorial Library Collection, Truman State University.

Chapter 4: **p. 53,** Reprinted, with permission, from *J. Arthur Harris. Botanist and biometrician* (ed. C. Otto Rosendahl et al.), 1936, University of Minnesota Press; **p. 55,** from "Charles W. Metz," 1922. History of the Marine Biological Laboratory (http://hpsrepository.mbl.edu/ handle/10776/2514). Marine Biological Laboratory Archives, Woods Hole, MA; **p. 56,** from Metz C. 1916. *The American Naturalist* 50: 587–599, published by The University of Chicago Press for The American Society of Naturalists; **p. 58,** from Blakeslee AF. 1922. *The American Naturalist 56:* 16-31, published by The University of Chicago Press for The American Society of Naturalists; **p. 60,** from Little CC. 1920. *Science 51:* 467–468, American Association for the Advancement of Science (AAAS); **p. 63,** courtesy of American Philosophical Society; **p. 66,** From Blakeslee AF, Belling J. 1924–1925. *J Hered 15:* 195–206, © Oxford University Press; **p. 67,** from Blakeslee AF. 1922. *The American Naturalist 56:* 16-31, published by The University of Chicago Press for The American Society of Naturalists.

Chapter 5: **p. 73, left,** Courtesy of The Suffolk County Vanderbilt Museum, Centerport, New York; **p. 73, right,** courtesy of the Society for the Preservation of Long Island Antiquities; **p. 73, bottom,** DeGolyer Library, Robert Yarnall Ritchie Photograph Collection, Southern Methodist University.

Chapter 6: **p. 85,** Reprinted by permission from Macmillan Publishers Ltd, *Nature* 134: 102–103, copyright © 1934; **p. 86, bottom,** from *Evening Public Ledger* (Philadelphia, PA), December 5, 1930.

Chapter 7: **p. 94, p. 95,** Courtesy of Jan Witkowski.

Chapter 8: **p. 110,** Courtesy of American Philosophical Society; **p. 113,** from Painter TS. 1934. *Genetics* 19: 175–188; **p. 114, top,** from *Modern Mechanix Hobbies and Inventions Magazine,* Vol. XVI (16), No. 4 (August 1936); **p. 115, top,** From Bridges, CB. 1935. *J Hered* 26: 60–64, © American Genetic Association, reprinted with permission; **p. 122, top,** from Blakeslee A. 1932. *Proc Natl Acad Sci* 18: 120–130; **p. 123,** from Riddle O, Braucher PF. 1931. *Am J Physiol* 97: 617–625; **p. 125,** from Shannon C. 1940. "An algebra for theoretical genetics." PhD thesis (unpubl.).

Chapter 9: **p. 136,** U.S. Patent Office, July 27, 1948; **p. 144, top,** Neel JV. 1943. *J Hered* 34: 93–96 (copyright © 1943, Oxford University Press); **p. 145, top,** courtesy of Otis Historical Archives, National Museum of Health and Medicine.

Chapter 10: **p. 149,** Collection of the Huntington Historical Society, Long Island, NY; **p. 154, top,** Nordisk Pressefoto, Copenhagen, Denmark; **p. 155,** CSHL Archives, courtesy of The Delbrück Family; **p. 159,** © Research and Development Division, Schenley Laboratories, Inc., Lawrenceburg, IN; **p. 160,** reprinted from Kelner A. 1949. *Proc Natl Acad Sci* 35: 73–79; **p. 162,** reprinted by permission from Macmillan Publishers Ltd, *Nature* 163: 949–950, © 1949.

Chapter 11: **p. 171,** CSHL Archives, courtesy of The Delbrück Family; **p. 174, top,** Courtesy of Jan Witkowski; **p. 176,** courtesy of Esther M. Lederberg papers (SC0874), Dept. of Special Collections and University Archives, Stanford University Libraries, Stanford, California; **p. 185,** from Abramson HA, Evans LT. 1954. *Science* 120: 990–991, reprinted with permission from AAAS; **p. 186,** reprinted from Polner R. 1979. "P&S prof conducted LSD tests," **p. 1,** *Columbia Daily Spectator*, Vol. CIII, No. 111 (16 May 1979), © Spectator Publishing Company. All Rights Reserved.

Chapter 12: **p. 193,** Courtesy MIT Museum; **p. 196,** National Academy of Sciences—National Research Council, Washington; **p. 198, top,** reproduced with permission of the Barbara McClintock Papers, American Philosophical Society; **p. 201,** courtesy of *Long Islander News*.

Chapter 13: **p. 216,** Courtesy of the James D. Watson Collection, CSHL Archives; **p. 220, top,** reprinted from Cairns J. 1966. *J Mol*

Biol 15(1): 372–373, © 1966, published by Elsevier Ltd; **p. 220, bottom, p. 221, top,** courtesy of Ann M. Skalka; **p. 228, bottom,** courtesy of Rockefeller Archive Center; **p. 229, top, p. 230, p. 232,** courtesy of the James D. Watson Collection, CSHL Archives.

Chapter 14: **p. 235,** Courtesy of the James D. Watson Collection, CSHL Archives; **p. 237,** reprinted from Dulbecco R. 1952. *Proc Natl Acad Sci* 38: 747–752; **p. 239,** courtesy of the James D. Watson Collection, CSHL Archives; **p. 241,** reprinted from the *Journal of Molecular Biology* 70: 57–71, © 1972, with permission from Elsevier; **p. 242, top,** reprinted with permission from *Biochemistry* 1973, 12(16), 3055–3063, © 1973 American Chemical Society; **p. 243,** reprinted with permission from *Biochemistry,* 1973, 12 (16), 3055–3063, © 1973 American Chemical Society; **p. 244,** National Cancer Institute (Linda Bartlett, photographer); **p. 247, top,** reprinted from Lazarides E, Weber K. 1974. *Proc Natl Acad Sci* 71: 2268–2272; **p. 249, top,** courtesy of John Cairns; **p. 249, bottom,** photograph by Rick Stafford (used with kind permission of his estate), published in *Harvard Magazine,* October 1976; **p. 250,** reprinted from *Cell* 8, 163–182 © 1976, with permission from Elsevier; **p. 252,** reprinted from *Cell* 12, 1–8, © 1977, with permission from Elsevier; **p. 253,** with kind permission of Anne Meier.

Chapter 15: **p. 259,** Courtesy of Cold Spring Harbor Laboratory Press; **p. 272,** courtesy of Cold Spring Harbor Laboratory Meetings & Courses; illustration by Jim Duffy; **p. 287,** courtesy of CSHL Archives (photo by Norman McGrath); **p. 290,** courtesy of NIH/NCI.

Chapter 16: **p. 297,** Photo by Jan Witkowski; **p. 302, p. 305,** courtesy of CSHL Archives, Banbury Center Collection; **p. 306,** courtesy of Jim Hope; **p. 310,** TT News Agency/Sipa USA; **p. 312,** from Roberts L. 1991. *Science* 254: 184–186, DOI:10.1126/science.1925568, reprinted with permission from AAAS; **p. 313,** courtesy of the James D. Watson Collection, Cold Spring Harbor Laboratory Archives.

Chapter 17: **p. 319, left,** Wikipedia Commons, by User:Sanfranman59/NRHP California Sonoma; **p. 319, right,** photo by Jan Witkowski; **p. 320,** courtesy of Jim Childress, Centerbrook Architects and Planners, LLP; **p. 323,** photo by Jan Witkowski; **p. 326, top,** photo by Jan Witkowski; **p. 326, bottom,** Reprinted from Stewart DJ. 2000. *Genome Res* 10: 1–3, with permission of Cold Spring Harbor Laboratory Press; **p. 331,** courtesy of SUNY Stony Brook University; **p. 338,** © Centerbrook Architects and Planners, LLP; **p. 339,** photos by Jan Witkowski; **p. 345, top,** Bush V. 1945. Science, the endless frontier: A report to the president. U.S. Government Printing Office, Washington, DC; **p. 346,** photo by Jan Witkowski.

Index

Page references in *italics* refer to information found in the figures and their legends.